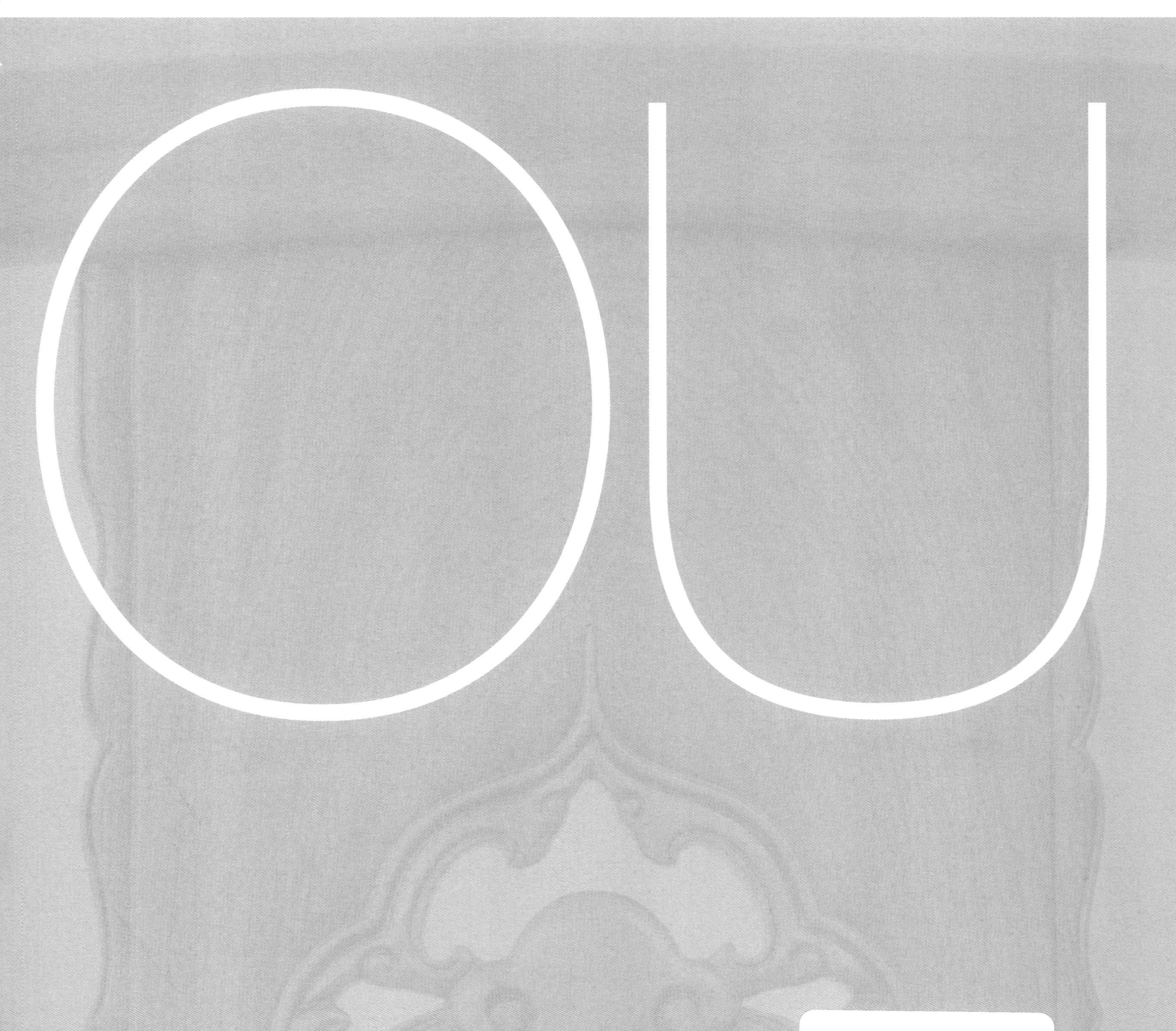

U0436509

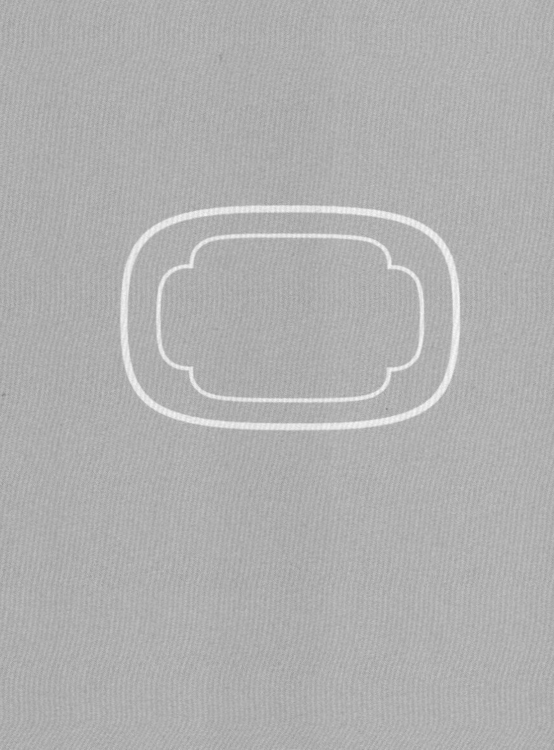

FURNITURE COLLECTION OF FAMILY OU

VOLUME I

Compiled by Ou Shengchun
Published by Zhonghua Book Company

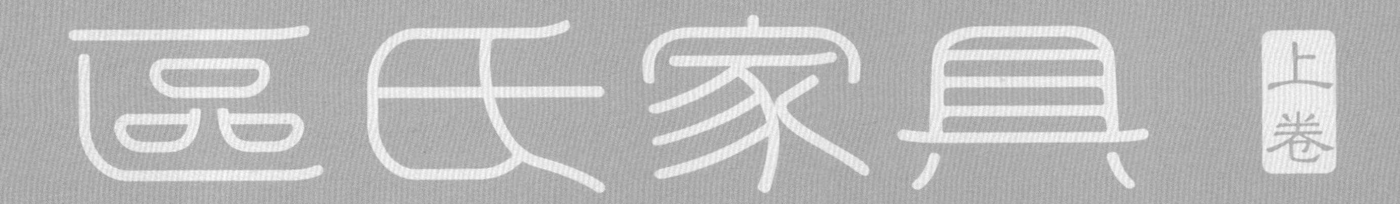

区胜春 著

中华书局

时间的居所

如何居住，这是人们与生俱来的情结。

安身才能立命，人们不仅仅满足于一个遮风避雨的空间，一间理想的居所应能表达出主人的趣味与心灵，也能看出主人对生活的努力与探索。个人品味在家居中有着重要作用，物主的教养和志趣很容易从中得到辨别，家是一个最诚实的地方，澄澈如水，映射出物主生活方式的选择与精神境界的追求。

时间没有尽头，一刻也不停息。人们总是希望能将时间的砖块砌入家的空间，获得更长的生命体验。晚明文人追求室庐“宁古勿时”“古雅可爱”，文艺思潮以复古为时尚。尚古不仅是人们对于文化溯源的坚守，更是对时间的无尽贪想，居所中陈列的古雅器具是往昔岁月的视觉化重现，“时间”因此能看得见、摸得着。尚古并非文化的一成不变，而是提供一种思考过去的方式，去芜存菁，选择前人创造的精华继续传递，完美当下的生活。

居所中最重要的陈设是家具，家具在室内构建起特定的文化情境，体现出人类的生活习惯和心理依归。中国有非常悠久的家具制作历史，不断传承与精进，最终在明清时期造就中国家具的黄金时代。中国古典家具是古代木作智慧和工艺的集大成者，东方的审美趣味能符合现代人的需求，北欧现代家具设计亦从中国古典家具中汲取了许多营养。Old is new，经过时间的洗礼，经典可能包含着最前卫、最时尚的内涵。

苏东坡曾为椅子赋诗一首：“无事此静坐，一日似两日。若活七十年，便是百四十。”舒适美观的家具能让人产生时间停滞的错觉，人、家具、时间嵌入空间的居所，目光可以驻留与凝视，内心达到安静与平和，人因此停留在家中的时间就更长了，心思也更远了。

世界是宽阔的，人们对远方永远有着无限的好奇与幻想。丝绸之路上蜿蜒的驼队，满载着波斯的金器和长安的丝绸，乾隆皇帝在圆明园里筑起西洋式建筑与喷泉，土耳其托普卡帕宫廷中摆满了青花瓷器——对异域文明的热切渴求是时尚的不变主题。1936年，多丽丝·杜克在夏威夷建造了一座传奇的居所，融合了美国东海岸、夏威夷和伊斯兰的建筑风格，房屋里摆满了来自印度、伊朗、叙利亚的华美地毯、陶罐和漆柜。大卫·纪德在京都的桃源洞居所将中国的家具和日本的建筑融为一体。安思远在纽约的大宅被公认为是东西方艺术的完美结合，也正是他的不懈努力，北美诸多重要博物馆有了中国家具的收藏。

身不能至，心向往之。面对家中陈设的异域美物，目光停留之处，远方变得触手可及，空间距离的拉近亦是时间的延长，时间与空间如此奇妙地互相转换，人们足不出户，神游千里。对家中的摆放器物细节充满爱意的掌控，也是为了找寻对人类世界之多样性的理解。

全球化的今天，世界趋向大同，而中国文化自有一份谦虚克制的内质，中国古典家具也有良好的适应性，自如融入不同的环境，相得益彰。

法国作家安德烈·纪德在《人间食粮》一书中写道：“你永远也无法理解，为了使自己对生活发生兴趣，我们曾付出了多大的努力。”居所是主人的知识积累、生活品味和生命追求的美学显现，因为对生活浓厚的兴趣而付出了巨大努力。王羲之曾言：“当其欣于所遇，暂得于己，快然自足，不知老之将至。”诚如是。

安思远在纽约的大宅
被公认为是东西方艺术的完美结合，
也正是他的不懈努力，
北美诸多重要博物馆有了中国家具的收藏。

Robert Hatfield Ellsworth's residence in New York is known to the world as a perfect fusion of eastern and western arts, and with his unremitting effort, many significant museums in the north America introduced Chinese furniture collections.

中山区氏家具有限公司始建于 1982 年，一直致力于传承中国传统家具文化与制作工艺，从未间断。四十多年的发展中，创始人区胜春坚守“还原经典”的企业理念，恪守产品的艺术品质，以材美工妍、雅正精湛的产品而享誉业界，深受人们的喜爱。家具是居室环境中最为重要的物品，也和人们的生活息息相关，作为全国性的传统家具制作品牌，我们热爱自己的工作，并将继续努力，希望能将更多家具精品带给人们，传递中国家具之美。

2012 年，区氏家具参与了中央电视台纪录片《家具里的中国》的拍摄工作；2013 年区氏家具荣获“2013 年度最具影响力的中国红木家具十大品牌”大奖；2015 年，在北京人民大会堂，区氏家具荣获“2014 年度最具影响力的中国红木家具十大品牌”大奖；2015 年 11 月，在北京钓鱼台国宾馆，区氏家具当选为全联艺术红木家具专委会主席团主席单位；2016 年 1 月，在海南博鳌，区氏家具荣获“中国红木家具四大艺术品牌”称号；2017 年 1 月，在杭州 G20 峰会会场，区氏家具蝉联“中国红木家具四大艺术品牌”称号。2019 年 6 月北京，佳士得艺术中心举办“汲古涵今”展览，首次将中国传统家具和现代艺术品融合在现代居室空间，获得了业内外广泛的赞誉。

改革开放以来，中国社会快速发展，变革迅猛，我们的生活日新月异，居住环境发生了巨大的改变。人们对美的生活怀着热切希望，一方面初步有了全球化的现代视野，一方面对自己的文化传统渐渐滋生兴趣，这样有趣的对撞既是新课题，也是我们的血脉基因。中国文化之所以可以一直不间断传承，就在于我们历史上不断地汲取外来文化，融入当下，并成为新的传统，生生不息。

The Dwelling of Time

How to reside, as an innate need and emotional consolation for human beings, has been contemplated throughout history of millenniums. As taught by Confucius, not until a man settles down under a roof on his own, would he be able to truly start a life. Ancient Chinese will not merely be content with a shelter space, but an ideal residence which reflects the owner's taste and spirit, hardworking and exploration for a better life. Personal taste plays an important role in the home design upon which owner's education background and aspiration can be easily demonstrated. Therefore, home, is a place where people can stay true to themselves, and is as clear as water, which can reflect the owner's choice of life style and spiritual pursuit.

Time flees constantly and unceasingly. People always wish to encapsulate each moment of life in bricks and stones for their home building, so as to acquire a longer experience for life. Litterateurs of the late Ming Dynasty (1368-1644) pursued to decorate home with antiquity; while in literature and art, the revisit of vintage style also prevailed. To uphold the classics isn't merely an inheritance and protection of origins of Chinese culture, also an endless covetousness for time, as those ancient objects displayed in the residence are visual recurrence of old ages, time hereby becomes visible and tangible; it isn't necessarily unalterable, rather than offering an approach to contemplate the past, to distinguish the refined essence and carry out the spirit of ancestors to perfect the present life.

Furniture is of the most significance in interior design, as it can construct a specific cultural context and mirror owner's life style and spiritual home. The history of furniture-making stretches several thousand years in China and culminates in its golden age during the Ming Dynasty and Qing Dynasty (1644-1911). The ancient Chinese furniture epitomizes the crowning designs and craftsmanship of ancient woodwork. The oriental aesthetics surprisingly satisfies the demand of modern societies, for instance, Nordic furniture design draws a lot from classical Chinese furniture. "Old is new", with time, the classics might contain the most progressive and fashionable connotations.

Su Shi (Su Tungpo, 1037-1101, a literary giant in the Northern Song Dynasty) once wrote a poem for a chair as "sitting on the chair with leisure and joy, seemingly to have lived two days within one; so if I can live to seventy years, then I should feel like twice of that age". Comfortable and beautiful furniture seems to be able to stop the time. People, furniture and time are embedded in the dwelling of space where the soul could stop and gaze in and mind comforted and pacified. People therefore would love to spend more time at home and their thought could stray further away.

The world is vast and broad, and people will always yearn for the yonder distant places with infinite curiosity and fantasy. The wandering camel caravan on Silk Road loaded with Persian gold-wares and Chang'an silk, western buildings and fountains built in Emperor Qianlong's Old Summer Palace, blue-and-white porcelains displayed in Turkey's Topkapi Palace, all demonstrate the keen desire for exotic culture is the constant theme of fashion. In 1936, Doris Duke built herself a legendary residence in Hawaii. As a fusion of American east coast, Hawaii and Islamic architectural styles, all the rooms were filled with gorgeous carpets, pottery pots and lacquer cabinets from India, Iran and Syria; David Gide decorates Japanese architecture with Chinese furniture in his Peach Blossom Abode in Kyoto; Robert Hatfield Ellsworth's residence in New York is known to the world as a perfect fusion of eastern and western arts, and with his unremitting effort, many significant museums in the north America introduced Chinese furniture collections.

Although ancient ancestors couldn't travel through time and space, they still longed for the distant. With the presence of various exotic objects at home, these antiquities seemed to draw the faraway places near to the front of beholder's eyes whenever people laid eyes on them. With spaces encapsulated, the time seems also expanded. Such miraculous shifts and transformations of time and space enabled people to travel through history across the map in their minds without setting a foot outside their houses. With lots of attention to detail and meticulous care and love, those ornaments and antique objects definitely reflect the owner's understanding on aesthetic multi-formity across the world.

In today's globalization, the world tends to be unified, and Chinese culture has a unique inherent quality of modesty and restraint, hence classical Chinese furniture can also better adapt to different environments and complement each other.

French writer André Gide wrote in his book *The Fruits of the Earth (Les Nourritures Terrestres)*, "You will never know the efforts it cost us to become interested in life." Residence is the aesthetic representation of the owner's depth and scope of knowledge, taste and pursuit of life, which reflects his/her huge effort cost due to enthusiasm for life. Wang Xizhi, great master calligrapher in the Eastern Jin Dynasty (317-420), wrote in *Preface for the Orchid Pavilion*, "while encountering with something worth our passion, we shall feel so content that we even forget the passage of time."

Ou's Furniture Company (Zhongshan City, Guangdong Province), founded in 1982, has been committed to inheriting the culture and craftsmanship of traditional Chinese furniture, unremittingly, all the time. During its almost forty years' development, the founder, Mr. Ou Shengchun has been adhering to the philosophy of "restoring classics", strictly maintains a high artistic quality. His products are renowned to the industry with beautiful and high quality material and craftsmanship, elegant and exquisite appearance, hence widely pursued by the public. Furniture, as an important element in interior design, also deeply involves in residents' lives. As a nation-wide traditional furniture manufacturer, we have been devoted to our work and will continue to explore and strive to bring more classical furniture of high quality, to pass on the aesthetics of Chinese furniture.

In 2012, Ou's Furniture participated in the filming of CCTV documentary *China in Furniture*; In 2013, Ou's Furniture was endowed the honor of "Top Ten Most Influential Chinese Redwood Furniture Brands in the Year of 2013"; In 2015, at the Great Hall of the People in Beijing, Ou's Furniture was endowed the honor of "Top Ten Most Influential Chinese Redwood Furniture Brands in the Year of 2014"; In November 2015, at Diaoyutai State Guesthouse in Beijing, Ou's Furniture was elected as the chairman of the presidium of special committee of national artistic redwood furniture; In January 2016, at Bo'ao in Hainan Province, Ou's Furniture was rewarded with the title of "China's Four Great Art Brands of Redwood Furniture"; In January 2017, at the conference hall where G20 Summit was held in Hangzhou City, Ou's Furniture was re-elected with the title of "China's Four Great Art Brands of Redwood Furniture". In June 2019, Christie's Art Centre (Beijing) held a featured exhibition "Succession" for Ou's Furniture, for the first time juxtaposed traditional Chinese furniture with modern arts in a living space of present time, which was critically acclaimed within and outside the industry.

Since China's reform and opening-up policy was implemented in 1978, the country has been on the track of rapid development and profound transformation. Huge progresses are being made on daily basis, and people's living condition is also undergoing great changes. With ardent hope for a better life, Chinese start to broaden the horizon and obtain a global perspective, while also become more and more interested in our own culture. Such clash of cultures emerges as an intriguing topic, as well as a genetic mark carved in our blood. With us continuously absorbing foreign cultures, subtly implementing them in present society and transforming them into new traditions, Chinese culture can be constantly enriched and passed on through generations with endless vitality.

目录

上卷 VOLUME I

下卷 VOLUME II

图版目录

Plates

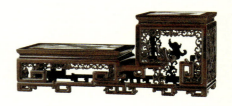

坐具

Furniture-to-sit-in

中国家具的历史，主要由席地而坐的低坐系统和垂足而坐的高坐系统构成，今天我们所能见的古代家具，基本都属于高坐系统。高坐系统约滥觞于汉代，唐代已渐成主流，五代、宋时已完全取代低坐系统。低坐系统的坐具以席和矮榻为代表；高坐系统的坐具极为丰富，有各式的凳、墩、椅、宝座等，甚至榻、罗汉床都可纳入坐具系统，可谓品类繁多，造型复杂。坐具最基本的构成因素是可以提供一个平面，或板面、席面，可供人坐于其上，另设四足，或有靠背、扶手等辅助性构造，整体造型比例适宜，加以多变的直线和曲线造型，颇有雕塑感，是家具中最受青睐的品种之一，尤其是明式坐具，倚坐舒适，且与现代家居环境可完美搭配，随意摆放，无不精彩。

The history of Chinese furniture is mainly constituted by low-seat system (with lower legs folded) and high-seat system (with lower legs stretched). And today's most existing ancient furniture belong to high-seat system which first appeared in the Han Dynasty (206 B.C.-220 A.D.) and gradually prevailed in the Tang Dynasty (618-907), then completely replaced low-seat furniture during Five Dynasties and Ten States (907-960) and Song Dynasty (960-1279). Low-seat system is represented by mat and short-footed couch; and high-seat system includes various kinds, like stools of multi-forms, footstool, chair, throne, etc., even couch and arhat bed can be included as well. With assorted varieties and intricate forms and designs, all these furniture constitute the whole gorgeous and exquisite seating furniture system. It basically provides a flat or plank or mat for people to sit on, with four legs, sometimes a back, armrests, etc. to supplement the function, presenting a proper structural proportion on the whole, adding variable straight and curve lines to conjure up a strong sense of sculptural beauty. It's undoubtedly one of the most acclaimed kind of furniture in China, especially seating furniture of Ming Dynasty which are comfortable and can seamlessly fit in the modern interior design.

凳

Stool

凳原写作“櫈”“橙”等，至迟魏晋时已有，敦煌莫高窟257窟北魏壁画中就能见到。凳的造型，起初甚是简单，一板植以四足即是，这种简练的造型绵延千余年，今日依然。同时，随着中国家具的发展、成熟，凳的造型也日益丰富，可简可繁，凳面除了常见的方形、长方形、圆形外，还有委角方形、海棠形、梅花形、菊瓣形、腰圆形、葵花形等，腿足也有三足、四足、五足乃至多足，尺寸可大可小，小者十余厘米，携带使用方便，大者七十厘米甚至更多，名为“禅凳”，可便跏趺坐参禅。在现代家居生活中，凳可随意摆放，造型活泼多变，无所不可，既可充坐具之用，视需求而增减，也可置于一旁，搁置物品，作矮几之用。

Stool (凳), originally written as “ 櫈 ” or “ 橙 ”, appeared in China no later than Kingdom of Wei (220-265) and Jin Dynasty (265-420), and was illustrated in the mural paintings of Mogao Grottoes (cave 257, the Northern Wei Dynasty, 386-534). It begins with a simple structure of having one seat with four legs, and still maintains the design up till present through millenniums. Meanwhile, with the development and maturity of Chinese furniture, the appearance of stool becomes varied with simple or exquisite decorations, multiple shapes of the seat (including common square, rectangular and round panels, as well as dented-cornered square, begonia-shaped, cinquefoil-shaped, chrysanthemum-shaped, oval and sunflower-shaped, etc.), multi-numbered legs (including three-legged, four-legged, five-legged or multi-legged) and different sizes from the portable smallest of only a dozen centimeters tall, to the larger one of seventy centimeters or more in height, called Zen stool used during cross-legged meditation. In modern living space, stool can be placed or assembled with other furniture randomly. With easy and changeful forms, it can function differently and freely depending on people's need, sometimes as furniture to sit in or end table to display.

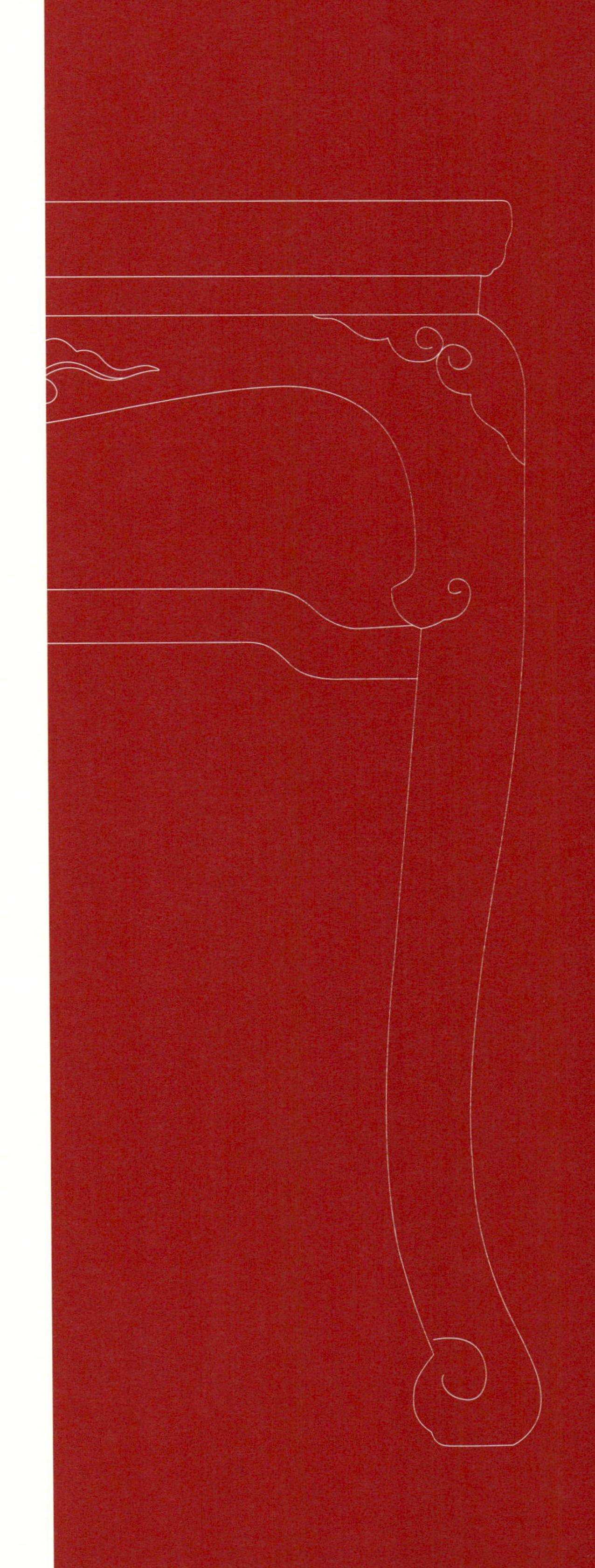

001

罗锅枨瓜棱腿方凳

Square Stool with Hunched Rails
and Grooved Legs of Melon-shaped Section

50×50×49cm

方凳由黄花梨制成，选料精良，纹理流畅，光泽莹润稳重。

座面为四面攒边，面心装软屉，编精致纹理的席面。边抹上舒下敛，中起线，成冰盘沿压边线。座面下按洼膛肚牙板，牙头与牙条相接处各透锼卷云一朵。牙板沿边起阳线，与腿、枨呼应。

座面下四腿八挓落地，腿足截面成内方外瓜棱状，压边线，腿足看面瓜棱凹入处起一炷香。此腿足截面形状处理与传统瓜棱腿做法不同，饱满多变，多有新意。

四腿之间，牙板之下按罗锅枨，罗锅枨截面劈料做法，上下各压边线，饱满圆浑的混面与腿足截面的瓜棱状有异曲同工之妙。

此件瓜棱腿方凳胜在腿足、罗锅枨上的线脚变化，统一中有变化，变而不乱，殊为难得。

It's made of yellow *Huali (Huanghuali)* wood of refined material with smooth veining and elegant gloss.

The stool has an assembled seat frame, a soft woven cushion, an exquisite rattan seat surface; Seat edges are narrowed down widthways (also called ice-plate edge) and planed down a rabbet in the middle and another one at the bottom rim as the borderline; under the seat panel, assembles four apron plates with embossed borderlines to echo the same design carved in legs and rails, and with hollowed cloud patterns at each corner.

Four splayed legs touch the ground with melon-shaped sections. Along the joint of apron plates and legs and between two melon arcs on the front side of legs, are embossed with a line, hence its melon-shaped leg section varies from other traditional kinds with fuller composition and more originality.

It has four hunched rails between four legs below apron plates, planed into two cambered surfaces on exterior with two rabbets along top and bottom rims. The round and full cambered surfaces visually resemble melon arcs along the legs.

This rare stool is quite distinguished with many intricate designs and changes in legs and hunched rails which vary from tradition yet are unified in one style.

002

卷草纹罗锅枨三弯腿长方凳

Rectangular Stool with Hunched Rails, Arabesque Pattern and Three-curved Legs (Cabriole Legs)

51×41×50cm

方凳由黄花梨制成，木纹细腻流畅，色泽莹润。

座面四面攒边，面心装软屉，席面编织精细。边抹上舒下敛，冰盘沿压边线。

座面下收矮束腰，下承三弯腿，腿足曲线不似一般的三弯腿外翻，而是向内收成卷云，内敛轻雅。束腰下腿足上肩膀处模仿铜包角的做法浅浮雕如意云头纹饰。

两腿之间做壶门牙板，牙板上浅浮雕卷草纹饰，卷草翻转舒展自如。牙板下，两腿之间装素罗锅枨，腿足在近罗锅枨相接处内翻一小云头，以呼应罗锅枨。从结构上考虑，出这一小云头可以增加腿足的强度，为罗锅枨出榫提供坚固的基础，这一处显示出造型与结构的完美融合。座面之下的壶门牙板和腿足起阳线一圈，强调优雅的曲线轮廓。

此件方凳装饰得体，恰到好处，格调清雅秀丽。

原物载王世襄《明式家具珍赏》。

It's made of yellow *Huali (Huanghuali)* wood of refined material with smooth veining and elegant gloss.

The stool has an assembled seat frame, a soft woven cushion, an exquisite rattan seat surface; Seat edges are narrowed down widthways (also called ice-plate edge) and planed down a rabbet at the bottom rim as the borderline.

Under the seat panel, it has a short girdled waist directly connecting with four three-curved legs (cabriole legs). Legs are curved elegantly inward as cirrus cloud instead of outward in general tradition. At tips of four shoulders of legs are decorated with *Ruyi*-shaped cloud bas-relief mimicking copper corner protectors.

Between legs, apron plates are carved out with decoration of smooth and graceful arabesque patterns; below apron plates, plain hunched rails are installed between legs; a small cloud-shaped bas-relief is carved out near each joint of legs and hunched rails to echo four rails. Structurally, the small cloud bas-relief can enhance the mechanical strength of legs and provide a solid base to take in tenons of hunched rails, hence a perfect combination of form and structure. Rims of apron plates and legs are embossed with a line to stress its elegant curve contour.

This square stool is of proper decoration and exquisite detail.

Reference to *Classic Chinese Furniture: Ming and Early Qing Dynasties* by Wang Shixiang, page 63.

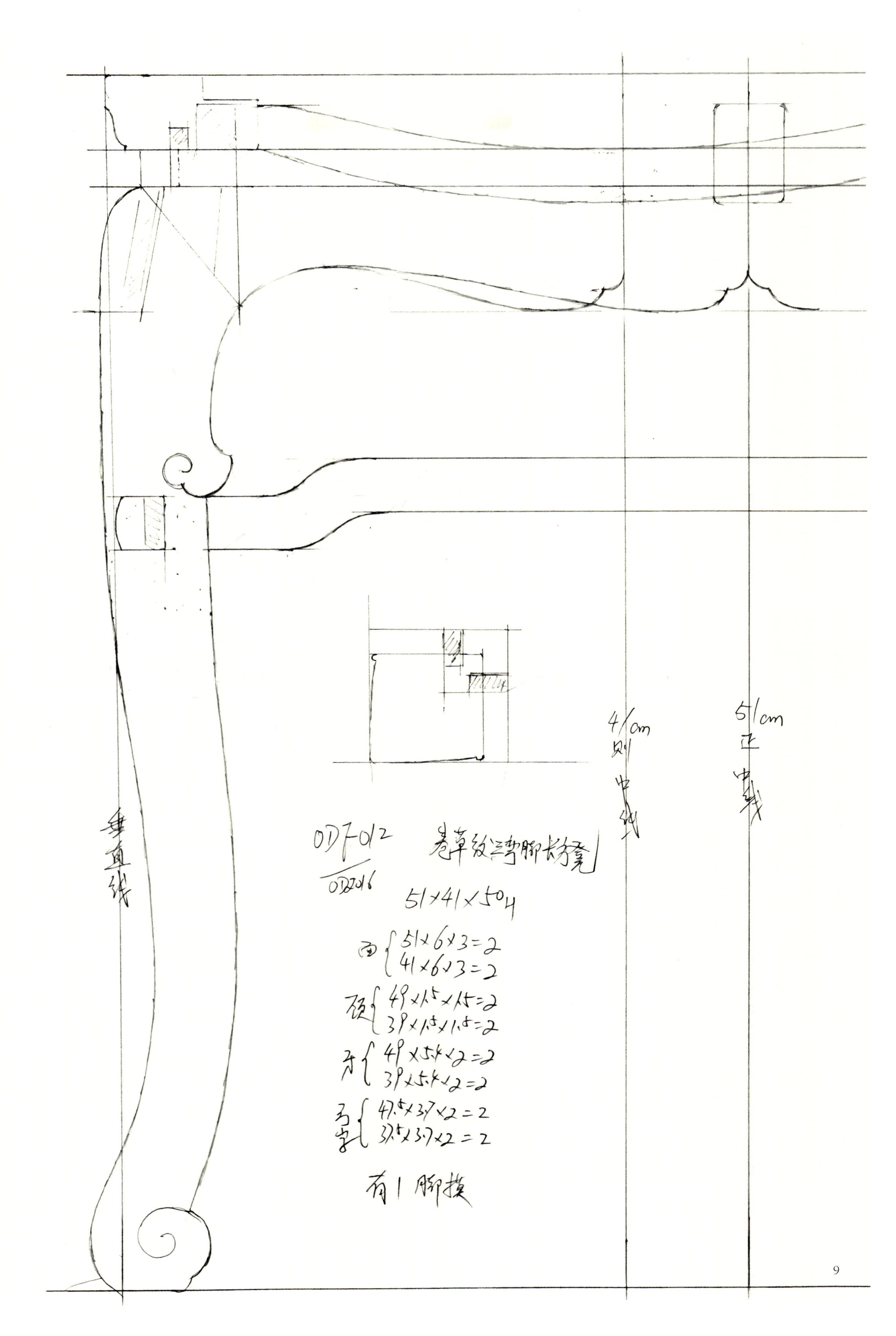

41cm
侧中线
51cm
正中线
垂直线
ODF-012
ODZ016
卷草纹三弯脚长方凳
51×41×50.4
面
51×6×3=2
41×6×3=2
49×1.5×1.5=2
39×1.5×1.5=2
牙
49×5.5×2=2
39×5.5×2=2
47.5×3.7×2=2
37.5×3.7×2=2

003

霸王枨打洼委角方凳

Square Stool with Curved Diagonal Beams
and Chamfered Dented-corners

41×41×44.5cm

方凳由白酸枝料制成，选料精良，木纹细腻，色彩莹润。

方凳采用四面平的做法，不同之处在于横竖材相交处作圆角处理，留出圆润混面。

座面四面攒边，面心装软屉，席面编织。边抹作打洼委角处理，转角下去正与腿足截面的双面打洼委角处理相联通，将打洼委角向下延伸至腿足下的管脚枨，一气呵成，形成此件方凳最独特的造型处理。

座面底部装霸王枨，霸王枨挺拔向上，力显气宇轩昂之势。

腿足下端装管脚枨，与座面用料、做法皆相同，成上下对称的长方造型。托泥之下安四小足，将托泥隔离地面，隔湿耐潮。

此件方凳以横竖方材围合成一个长方空间，各方材线脚皆作打洼委角处理，既简洁又精致，八面玲珑，有明显的立体雕塑感，更是造型与结构的完美结合。

It's made of white dalbergia cochinchinensis of refined material with smooth veining and elegant gloss.

The stool is with flat-corner structure (corners butted with mitered corner bridle joints), with joints of vertical and horizontal frames carved into curves.

The stool has an assembled seat frame, a soft woven cushion, an exquisite rattan seat surface; vertical sides of seat edges are planed down into low-lying surfaces, so are exterior sides of four legs and four rails; and eight tips of corners are chamfered to form a unified style which is the most unique design of this stool. Under the seat, four curved diagonal beams are raised up to show strength and beauty of mechanics.

Below four legs connects four foot rails made of same material and processed with same craftsmanship to form a rectangular frame with symmetrical design. Four small foot pads are installed at bottom corners of foot rails to isolate the stool from ground to resist humidity.

This stool is assembled with square-sectioned materials to frame a rectangular space; each material is planed down and chamfered at corners, so as to make the stool an elegant and delicate sculpture of geometrical beauty, as well as a perfect combination of form and structure.

004

带托泥委角方凳

Square Stool with Foot Rails and Dented-Corners

39×39×48cm

方凳由白酸枝料制成。

方凳座面为正方委角，四面攒边，格角榫采用闷榫，不留榫头于器表，使得整器更加美观。面心装软屉，席面编织。边抹作泥鳅背双面压边线，混面圆润饱满。座面下装矮束腰，亦委角，与座面委角相呼应。

束腰下装素直牙条，四角装C形弯腿，腿截面内方外圆，外部圆混曲线正中委角，与座面和束腰的委角相呼应，向下延伸至足下。腿足下端向两侧翻转成云头收尾。

四足踩托泥，托泥四角亦委角，边抹与座面边抹做法相同，作泥鳅背双面压边线，下承小足，将托泥与地面隔离，防潮隔湿，又增加足下空间的空灵感。

整件方凳胜在空灵秀雅，以线材构件勾勒出虚空间，委角装饰从座面一直延伸至小足，自成一体。

It's made of white dalbergia cochinchinensis.

The square seat panel is with four dented corners and assembled frames butted with hidden dowels to maintain a clean and beautiful appearance; and it has a soft woven cushion, an exquisite rattan seat surface; four edges of the seat are polished in loach-back style and with two rabbets on both sides as borderlines; all cambered surfaces are full and smooth. Under the seat, it has a short girdled waist also dented at four corners to echo the seat panel. Below the waist installs plain apron plates and four c-shaped legs of half-square-and-half-round section with their exterior cambered surfaces dented in the middle stretching along four legs all the way to the bottom of four feet to echo the seat and girdled waist; Tips of four feet are split sideways and rolled over as a cloud.

Four feet touch foot rails which are also dented at four corners. Edges of foot rails are processed with same craftsmanship as seat panel, polished in loach-back style and with two rabbets on both sides as borderlines. Small foot pads isolate the stool from the ground to resist humidity and also give it a sense of lightness. The stool is quite distinguished with its ethereal elegance, as all these components perfectly outline a void space. Dented-corners are applied on the seat panel down to foot rails as a unified style.

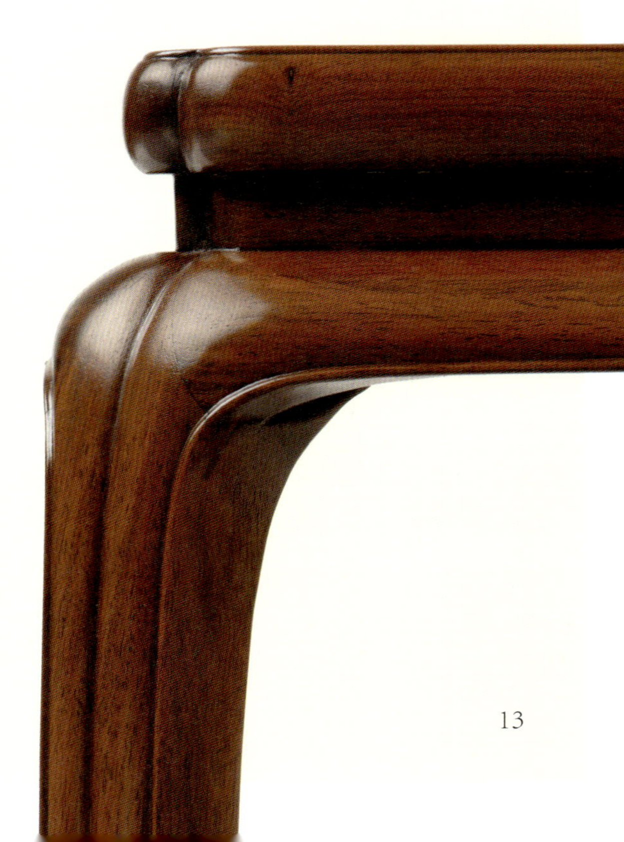

005

带托泥五卷足圆凳

Round Stool with Five Curved Feet and Foot Rails

38×38×50cm

圆凳由红酸枝料制作，纹理内敛，色泽沉穆。

座面圆形，面心硬屉，踩鼓落膛心。边抹作泥鳅背双面压边线，混面圆润饱满。座面下收束腰，束腰打洼。

束腰下装内卷五足，用料硕大，却大挖曲线，勾勒出秀美纤长的腿足曲线。两腿之间按洼膛肚牙板，牙板和五腿边缘皆起阳线，顺势而下，沿着内卷足部收尾。

五腿之下承圆形托泥，边抹亦作泥鳅背双面压边线，下承相应小足。托泥尺度和线脚处理与座面相似，只不按心板，留出空间，与座面既呼应，又有变化，使整件家具上下呼应，成完整一体。

此圆凳以流转曲线勾勒结构构件，形成饱满通透的空间，内卷五腿曲线遒劲中带柔美，似书法中的顿挫笔意，是此件作品的点睛之处。

明万历
《御世仁风》插图
凤阳刊本

Wanli Period of Ming Dynasty (1573-1620)
Illustration of *Collected Stories of Emperors*,
Fengyang edition

It's made of red dalbergia cochinchinensis with dark and subtle veining and solemn gloss.

Round seat panel with a hard cushion has a circular groove carved in the seat surface; edges are polished in loach-back style, with two rabbets on both sides as borderlines and a polished cambered surface; below the seat connects a short girdled waist caved in as a low-lying groove.

Below the waist connects five inwardly curved legs carved out of massive solid material to outline a slender and graceful silhouette. Between legs, it installs hollowed apron plates with embossed lines running along rims and connecting same bas-relief lines along rims of five legs, and reaching its end at tips of five rolled-up feet.

A circular foot rail is attached below five legs, with its edge as well polished in loach-back style with two rabbets on both sides, and with five small foot pads installed below the rail. The foot rail is processed with same craftsmanship as seat panel except that the centre panel is left out to make more breathing space, so as to echo the seat panel with variations and to complete the stool with a unified style.

The round stool is outlined with floating curved components to form a full and ethereal inner space; the five rolled-up legs are of vibrant and graceful beauty, as strokes of pause and turn in Chinese calligraphy.

椅

Chair

“椅”本指树木，椅子原作“倚子”，意为有后背可倚靠的坐具，后渐称为“椅子”。唐代开始出现“倚子”这个名词，但椅子的形象，在敦煌莫高窟285窟西魏壁画中已有所见，概其为僧侣禅坐的“绳床”，是随着佛教一起传入我国的，魏晋以来的壁画或佛典中常有所及。但日常生活中椅子的使用，当在唐中晚期，宋代时方大肆流行。椅子有交椅、圈椅、灯挂椅、靠背椅、四出头官帽椅、南官帽椅、两出头官帽椅、玫瑰椅、禅椅等。不仅品类丰富，而且造型优美，富于空间变化，是家具中最有雕塑美的品类，且可自由组合，单搁、对置、群放，无一不可，无一不美，既是不可或缺的实用器具，又是富于审美情趣的艺术品。东坡有“无事此静坐，一日似两日。若活七十年，便是百四十”句，周天球书而刻于椅背，将中国椅具造型意境之娴雅幽静，显现无遗。

Chair (椅), originally indicates the tree, formerly written as “倚子”, meaning a seating furniture with a back to lean on, and later is called “椅子”. According to existing records, the word “倚子” first appeared in Tang Dynasty (618-907), yet it had been already illustrated in mural paintings of Mogao Grottoes (cave 285, Western Wei Dynasty, 535-556) as a rope bed used by monks during their meditation, so it was probably introduced to China along with Buddhism. Ever since Kingdom of Wei (220-265) and Jin Dynasty (265-420), chairs are frequently referred in mural paintings and Buddhist scriptures; however, they weren't widely used in daily life until mid-and-late Tang Dynasty, and later prevailed across country in the Song Dynasty (960-1279). It includes folding chair, horseshoe-back armchair, yoke (lamp hanger)-back chair, backed chair, yoke (official's hat)-back armchair with four protruding ends, yoke (official's hat)-back armchair, yoke (official's hat)-back armchair with two protruding ends, low-back armchair and Zen chair (for meditation) etc., with wide range of variety and appealing designs of sculptural beauty, so that they can fit in any given situations and spaces, and can be freely re-grouped with other furniture, be placed in set, in pair or solely as an absolute artwork of indispensable function and aesthetic value. Su Shi once wrote a poem for chair as “sitting on the chair with leisure and joy, seemingly to have lived two days within one; so if I can live to seventy years, then I should feel like twice of that age”. And Zhou Tianqiu (famous painter and calligrapher in Ming Dynasty,1514-1595) had his calligraphy carved on the back of his chair, giving the furniture a tranquil and graceful quality.

006

素灯挂椅

Yoke (Lamp Hanger)-back Chair

49×43×48/104.5cm

灯挂椅由花梨木制成，精心选料，纹理优美细腻，颜色淡雅，有浓郁的文人书卷气息。

搭脑为一块木料挖成，中间宽阔，以承接头部，下接靠背，搭脑两侧舒展上扬，柔婉提气。靠背精选上好独板大料，成S形，与弯曲的人类脊柱曲线相合，增加舒适度，符合人机尺度。更可贵的是，独板背靠上的纹理更是优美，圈圈年轮以靠背偏上，黄金分割点的位置为中心，自然荡开，成为整件家具的视觉中心。

座面四面攒边，面心装软屉。边抹为冰盘沿压边线，干净利索。座面下装四腿，腿部截面外圆内方，后腿向上延伸，连接搭脑。

看面两腿之间按直牙条券口牙子，侧面两侧之间装刀板牙子。腿间皆装赶枨，两侧高，前后底，前面为榻脚枨。

此灯挂椅造型简洁洗练，毫无多余装饰，尽显秀雅。

明晚期
《新刻全像易鞋记》插图

Late Ming Dynasty (1368-1644)
Illustration of *Story of Shoe-changing*

It's made of yellow *Huali (Huanghuali)* wood of refined material with smooth veining and elegant colour, exhibiting a sense of literati aesthetics.

The headrest is carved out of a complete material with a flat and broad center part to hold up sitter's head, and is assembled to the below backrest plate, with its two ends protruding out and rolled up elegantly; the backrest is made of selected refined complete material, processed into ergonomic s-shape to perfectly fit the curvature of human spine to increase its comfortness. Moreover, as the visual focus of the backrest, subtle veining of captivating natural growth rings emerge at the upper part, the exact spot dividing the plate at golden ratio, and undulating ripples across the backrest.

The seat panel is constructed with assembled frames and a soft woven cushion; edges are narrowed down widthways (also called ice-plate edge) and planed down to two clean rabbets as borderlines; below the seat installs four legs with a half-round-and-half-square section; rear legs extend up and are directly butted into headrest rail.

Between legs on the front side installs square apron plates; and other apron plates on the side are knife-shaped; stretchers are assembled between legs: lower on the front and the back and higher on the side; the front stretcher is used as footrest.

This yoke-back chair epitomizes beauty of simplicity and elegance.

007

如意云纹开光灯挂椅

Yoke (Lamp Hanger)-back Chair with *Ruyi*-shaped Cloud Carving on Backrest

48×43×44.5/98cm

灯挂椅由红酸枝制成，选料整齐，色泽均匀，打磨精到，光泽如玉，气质内涵。

搭脑造型柔婉，牛头式，末端为鳝鱼头，含蓄内敛，给人以触之温润的感受。三弯形靠背板，在中间靠上位置设如意云头开光，并沿轮廓起线强调，是全器唯一的装饰之处，甚为醒目，宛若画龙点睛，令人第一眼就注意到这个精彩的装饰。后腿C形后弯，与靠背板形成势若拉弓的样式，充满了张力。

座面攒框装席面，即软屉做法，边框做成素混面，不起一线，与整体浑圆风格相合。座面下四面装刀牙板，沿边起线，简洁空灵。腿足外圆内方，下方设步步高赶枨，既有步步高升的良好寓意，又在视觉上形成层次变化，具空间感。前方横枨又名踏脚枨，较为扁阔，可以脚踩其上，下附牙条，以增牢固。

此灯挂椅为江南地区经典样式，所见榉木制者最为典型，今用硬木制成，形神亦佳。

It's made of finely-polished red dalbergia cochinchinensis of refined material with balanced colour and jade-like gloss.

The headrest is elegantly processed into ox-head shape with two ends polished in eel-head style, giving a subtle and feminine charm; On the upper side of the three-curve backrest is carved a hollow *Ruyi*-shaped cloud highlighted with embossed borderline as the only adornment on the chair that will attract beholder's eyes instantly; two c-shaped rear legs bent backward slightly as drew bows, gives a sense of strength and strain.

Seat panel has assembled frames and a soft woven cushion; Edges are flat and unadorned, to echo the general simple and well-rounded style; below the seat installs apron plates by T-bridle joints (clamp joint) on four sides with embossed borderlines; legs of half-round-and-half-square section are butted with stretchers to imply the wish of being promoted to a higher position and also to create a rhythmic change in spatial structure; the lower front stretcher is called footrest, with a flatter and broader side to support feet, and with apron plate installed below to reinforce its firmness.

Such kind of yoke-back chair represents the classical furniture widely used in the south of the lower reaches of the Yangtze river where chairs are typically made of beech. This one, made of hardwood, as well demonstrates high aesthetic value.

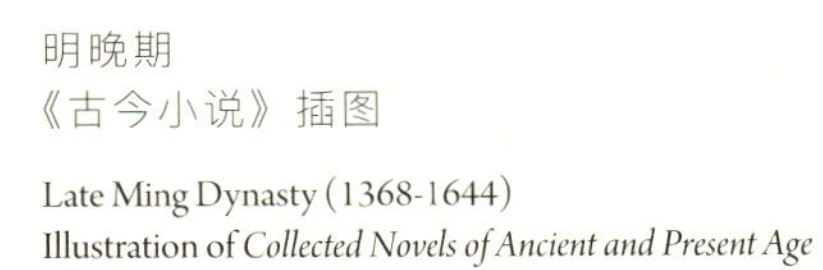

明晚期
《古今小说》插图

Late Ming Dynasty (1368-1644)
Illustration of *Collected Novels of Ancient and Present Age*

008

四出头官帽椅

Yoke (Official's Hat)-back Armchair with Four Protruding Ends

55×46×50/109cm

椅子由紫檀制成，紫檀颜色沉穆稳重，表面莹润。

搭脑由一块木材挖出颀长的曲线，曲线自中部承接靠背处徐徐外扬，自外端成鳝鱼头，起伏有势。

搭脑下精选独板靠背，上收下大，靠背纹理如山峦叠嶂，牛毛纹密密隐现。扶手成蜿蜒的S形，以鳝鱼头收尾，与搭脑呼应。扶手下鹅脖靠后安装，省去联帮棍。边抹做泥鳅背，不压边线，强调造型的干净利落。

座面四面攒边，面心装席面软屉。座面下接四腿八挓，腿截面内方外圆，体现中国"天圆地方"的哲学意境。看面两腿之间安素雅的刀牙板，四腿之间安步步高赶枨，有步步登高的美好寓意。

整件椅子素雅无饰，每一件构件都提炼至简，甚至不做多余的线脚处理，只让曲线游走，勾勒轮廓，充分体现了家具追求的雅致与精炼。

明万历
《古杂剧二十种》插图
顾曲斋刻本

Wanli Period of Ming Dynasty (1573-1620)
Illustration of *Twenty Classic Chinese Operas*, Guquzhai's edition

It's made of red sandalwood with dark and solemn colour and elegant gloss.

The headrest is carved out of a complete material with slender shape and a subtle curve in the middle where it connects to backrest, and two protruding ends rolled backward and polished in eel-head style; below the headrest it installs the backrest made of selected complete material, and slightly narrowed down on the upper side; it exhibits veining running as continuous mountains and subtle ox-hair veining (traces of capillary tubes of the plant); armrests are processed into s-shape with ends polished in eel-head style to echo the two ends of headrest; curved rods (also called "goose neck") below armrests lean backward to also function as supporting rods (joint rods) in the middle; edges are polished in loach-back style without borderlines to emphasize its clean outline.

Seat panel is constructed with assembled frames and a soft woven cushion; below the seat it installs four splayed legs of half-round-and-half-square sections to epitomize the philosophy of "round sky and square earth"; an unadorned apron plate is installed between two front legs by T-bridle joints (clamp joint); four stretchers are installed between legs to imply the wish of being promoted to a higher position.

This chair is unadorned, even without any extra borderlines, and assembled with highly reduced components only to leave curve lines running through to outline the silhouette, fully representing the aesthetics of simplicity and elegance.

009

四出头官帽椅

Yoke (Official's Hat)-back Armchair with Four Protruding Ends

54.5×45.5×50/110cm

椅子由白酸枝制成，精心选料，纹理流畅舒展，色泽雅致。

搭脑为一块整料斫制，勾勒出舒展上扬的曲线，增加了整件家具的气势。搭脑下精选纹理优美的独板做成靠背，纹理成涡漩状，一圈圈漾开，精美的木材自然纹理成为整件家具的视觉中心。扶手做优雅的S形曲线，收尾处外展，与搭脑收尾处呼应。扶手下接鹅脖，鹅脖后收，避开前腿上沿。

座面四面攒边，面心装席面软屉。边抹做泥鳅背，素混面饱满。座面下接八挓四腿，后腿上沿穿过座面上接搭脑。看面两腿之间安素刀板牙子，四腿之间安步步高赶枨，自看面、侧面、背面步步抬高。

整件家具素雅精炼，不着装饰，又善于表现木材的自然纹理，造型、结构和材料完美结合，有着浓郁的文人气息。

It's made of white dalbergia cochinchinensis of selected material with loose and smooth veining and elegant gloss.

The headrest is made of complete material to outline a rising curve on two ends, breathing a vigor to the furniture; below the headrest, it installs the backrest made of selected complete material with swirls of veining rippling outwards as the visual focus of the whole furniture; armrests are processed into s-shape with two ends slightly bent outward to echo two ends of headrest; below armrests connects curved rods (also called "goose neck") leaning backward to avoid joints of seat panel and two front legs.

Seat panel has assembled frames and a soft woven cushion; edges are polished in loach-back style with a fuller cambered surface; below the seat it connects to four splayed legs; two rear legs extend upward through seat panel to joint with headrest; between two front legs installs knife-shaped apron plates; four stretchers are installed at various height between four legs.

This chair is of simple and unadorned beauty to perfectly exhibit the natural veining of the material, as a combination of form, structure and material, reflecting the aesthetics of literati.

010

四出头官帽椅

Yoke (Official's Hat)-back Armchair with Four Protruding Ends

58×48×49/110cm

此椅黄花梨制成，与前件官帽椅造型相似，然又有不同的细节变化，前者空灵秀美如美人，此件落落大方如君子，正可对比品鉴。两者最大的区别在于靠背板的处理，前者 S 形三弯且较窄，得亭亭玉立之美，此件 C 形单弯且阔绰，得质朴素雅之美。古人云：“牵一发而动全身。”虽是一处细节的改变，其实搭脑、扶手、牙板等处曲线都随之变化，以取得整体和谐统一的效果，这正是匠师和设计者因地制宜，对造型语言融会贯通而成的佳作，也正体现了中国家具在造型变化和审美情趣方面卓越的成就。

It's made of yellow *Huali (Huanghuali)* wood with similar structure of previous piece, except for some variations in detail. The previous one appears like an ethereal beauty, while this one a decent gentleman. These two chairs are mostly distinguished in their backrests: the previous one is in s-shaped curve with narrower width; while this one is in c-shaped curve with wider width. As the ancients said "small detail can make big difference"; once backrests are altered, other parts like headrests, armrests, apron plates, etc., shall be adjusted as well to obtain a unified and harmonious style. These two pieces of furniture are perfect demonstration of craftsmen's improvision and designer's wisdom, as well as the aesthetic accomplishment and philosophy of Chinese furniture-making in form variations.

011

四出头官帽椅

Yoke (Official's Hat)-back Armchair with Four Protruding Ends

58×48×49/110cm

此件官帽椅与前件造型相同，但是材料不同，前者黄花梨，此为铁梨木，呈现出不同的质感，各有其美。

铁梨木是中国古代本土硬木，多产广西，有大料，纹路优美璀璨，色泽沉稳雅致，是性价比甚高的家具用材。以之所制明式家具，多有佳作。

This chair is of the same structure of the previous, aside from different materials (the previous is made of yellow *Huali* wood; this one lignumvitae) presenting different textures of unique beauty.

Lignumvitae is an ancient local hardwood in China, widely grows in Guangxi Province, with gorgeous veining and graceful gloss. It's a highly economic material and many classical Ming-styled furniture are made from it.

明崇祯
《金瓶梅》插图

Chongzhen Period of Ming Dynasty (1628-1644)
Illustration of *The Golden Lotus* (*Jin Ping Mei*)

012

罗锅搭脑四出头官帽椅

Yoke (Official's Hat)-back Armchair with Four Protruding Ends and Hunched Headrest

58×45.5×50/111cm

椅子由白酸枝制成，选料精良，着重表现木材的自然纹理。

搭脑用一整料挖出罗锅形，造型特别，与座面下的罗锅枨遥相呼应。搭脑下接独板靠背，靠背光素，尽显靠背自然的流畅纹理。靠背最下端开壶门亮角。扶手造型挺直，只在收尾处外展。扶手下接鹅脖。

座面四面攒边，面心装软屉，编席。边抹做泥鳅背素混面。座面下接四腿，腿截面内方外圆。座面下两腿间装罗锅枨，罗锅枨上装矮老与座面相接，形成错落的虚实空间。罗锅枨与矮老的组合，既是结构的需要，又是造型的塑造，实现了造型与结构的结合。四腿之间装步步高赶枨，既增加四腿之间的稳定性，又给人步步登高的美好寓意。

整件家具用料修长，多用直材，配合罗锅形曲线，勾勒轮廓，使得整件家具显得劲挺又空灵，气质独特。

It's made of white dalbergia cochinchinensis of selected material with natural veining.

The hunched headrest is carved out of a complete material with unique appearance to echo the hunched rail below the seat; below the headrest connects the unadorned backrest processed out of a complete material, fully showcasing the smooth veining of the natural beauty; at the bottom of the backrest, a hollow *Kun*-gate is carved out; armrests are straight and only bent outward a bit on the end; below armrests it connects rods ("goose neck").

Seat panel has assembled frames, soft woven cushion and a rattan surface; edges are polished in loach-back style; below the seat panel, it installs four legs of half-round-and-half-square sections; four hunched rails are installed between legs and reinforced with short rods; the combination of rails and rods is of mechanical function and also aesthetic expression; stretchers between legs increase the chair's steadiness and also imply the wish of being promoted to a higher position.

This furniture is made of slender and mostly straight components combined with curves of hunched rails to outline the silhouette, giving a unique upright solemnity and celestial beauty.

013

四出头官帽椅

Yoke (Official's Hat)-back Armchair with Four Protruding Ends

66×52×52/120.5cm

椅子由紫檀制成，色泽深穆，纹理流畅。

搭脑为一块木材斫成，曲线蜿蜒起伏，造型独特，搭脑横截面成方倒圆。搭脑下接独板靠背，靠背纹理如行云流水，棕眼丝丝隐现。扶手成S形，收尾处外展，横截面亦为方倒圆。扶手下接鹅脖。

座面四面攒边，面心装编席软屉。边抹做泥鳅背，混面无压边线。座面下接四腿，腿截面为方倒圆，与整件家具其他构件处理方法一致。看面两腿之间装直牙条券口牙子，不做线脚处理，任其光素。券口牙子下装踏脚枨，其他腿间安赶枨。

此件椅子最独特之处在于大部分构件横截面为方倒圆的处理，使得整件家具方正中显圆润，搭脑的造型处理也比较独特。

It's made of red sandalwood with dark and solemn colour and smooth veining.

The headrest is carved out of a complete material, with a unique undulating appearance and chamfered section; below the headrest connects an one-piece backrest with smooth veining, as well as subtle ox-hair (traces of capillary tubes of the plant) veining; armrests are s-shaped components of chamfered sections and slightly bent outward on ends; below armrests it installs curved rods ("goose neck").

The seat panel is constructed with assembled frames and a soft woven cushion; edges are polished in loach-back style without any borderlines; below the seat connects to four legs with chamfered sections to maintain a unified style; unadorned straight apron plates are installed between legs; below apron plates, one footrest are installed between front legs and three stretchers are connected between other legs.

Components of this chair (especially the headrest) are uniquely processed into chamfered section to add a sense of softness and gracefulness to its squarely solemn structure.

014

四出头官帽椅

Yoke (Official's Hat)-back Armchair with Four Protruding Ends

61.5×49×51/121cm

椅子由黄花梨制成，选材精良，纹理优美，色泽莹润。

搭脑为一块木材斫成颀长的造型，曲线优雅，中部与靠背连接处为曲线最高处，顺势而下，与后腿上沿构件连接处为最底处，然后顺势向后上方跳出，以鳝鱼头形收尾。

搭脑下接独板靠背，靠背精选出涡漩花纹，圈圈漾开。独板靠背上部与搭脑接触的位置，和下部与座面接触的位置各安角牙。上两角牙做云头曲线处理，外起阳线勾勒。下两角牙做曲线处理，外起阳线勾勒。扶手做S形曲线，与末端外展。扶手下装鹅脖和联帮棍。

座面四面攒边，面心装席面软屉。边抹做泥鳅背，上下各做压边线处理。座面之下接四腿，腿截面内方外圆压边线。看面两腿之间做壶门券口牙子，券口牙子下端做卷草曲线，与靠背角牙的处理呼应。券口牙子下装踏脚枨。其他腿间装赶枨。

整件椅子在精炼的造型基础上做了些许装饰，点到为止，恰到好处。

It's made of yellow *Huali (Huanghuali)* wood of refined material with smooth veining and elegant gloss.

The headrest is carved out of a complete material with slender form and elegant curve; it slightly rises up in the middle at the joint of headrest and backrest and then runs lower on two sides till it touches joints of two rear legs and headrest, and is finally bent upward on two eel-head polished ends.

Below the headrest, it installs an one-piece backrest with selected material and swirls of veining rippling across the plate; On upper and bottom ends of the backrest, corner apron plates outlined with embossed lines are installed and respectively carved into hollowed cloud pattern and curved design; armrests are processed into s-shape and slightly bent outward on two ends; and below armrests installs curved rods ("goose neck") and joint rods.

Seat panel has assembled frames and a soft woven cushion; edges are polished in loach-back style with two rabbets along top and bottom rims; below the seat it connects to four legs of half-round-and-half-square sections; apron plates of *Kun*-gate style are installed between two front legs, with bottom areas carved in arabesque design to echo the corner apron plates of the backrest; a footrest is installed between two front legs and three stretchers between other legs.

This chair is of concise form and adorned with highly abstract decorations to demonstrate the beauty of simplicity.

015

镶大理石三攒靠背四出头官帽椅

Yoke (Official's Hat)-back Armchair with Four Protruding Ends and Three-sectioned Assembled Backrest Embedded with Marble and Burl Plaques

60×48×49/116cm

椅子由紫檀镶大理石而成，选料精良，色泽沉穆稳重。

搭脑勾勒弯曲曲线，收尾处上扬，彰显气质。搭脑下接靠背，靠背三段攒成，上部镶嵌大理石，纹理似行云流水。中部镶嵌瘿木，纹理优美，水波盈盈；瘿木上浅刻。下部雕刻卷草云头纹亮脚。靠背两侧装整条云头曲线牙条，增添了几分精致。扶手做S形，于末端外展，以鳝鱼头形收尾，与搭脑收尾处呼应。靠背镶嵌材料的质地、纹理和色彩都呈现变化，加之适当的雕琢和曲线处理，使得整个靠背精致华美。

座面四面攒边，面心装硬屉。边抹做泥鳅背压边线，混面饱满。座面下四腿八挓，看面两腿之间安壸门卷草券口牙子，壸门牙条表面浅浮雕卷草纹，边缘起阳线以勾勒曲线。券口牙子下装踏脚枨。其他腿足之间装赶枨。

整件椅子以靠背的精致镶嵌与雕琢为视觉中心，其他构件一概精炼，简与繁的处理显实力。

This marble-embedded chair is made of refined red sandalwood with dark and solemn colour and elegant gloss.

The headrest is processed into undulating appearance with two ends slightly rolled upward to exhibit an aura of pride and solemnity; below the headrest, it installs the backrest assembled with three sections which respectively are embedded with a plaque of marble showcasing veining of floating cloud or water (upper), a plaque of burl with beautiful texture of rippling water and bas-relief (middle), and a corner plate carved in hollowed arabesque design (bottom); on two sides of backrest installs two one-piece straight apron plates carved in cloud pattern to enhance the delicacy of the furniture; armrests are processed into s-shape and slightly bent outward on two eel-head polished ends to echo two ends of headrest; the textures, veining and colours of the embedded material present different variations, with proper carvings and curvature process, highlighting the delicacy and extravagance of the chair.

The seat panel is constructed with assembled frames and a hard cushion; edges are polished in loach-back style and planed down two rabbets on both sides as borderlines; below the seat installs four splayed legs and arabesque apron plates of *Kun*-gate style carved with floral bas-relief and outlined with embossed borderlines; a footrest is installed between two front legs and three stretchers between other legs.

This chair is quite distinguished with the exquisite embedding and carving as the visual focus, while leaving other components simple and concise to demonstrate a perfect balance of simplicity and complication.

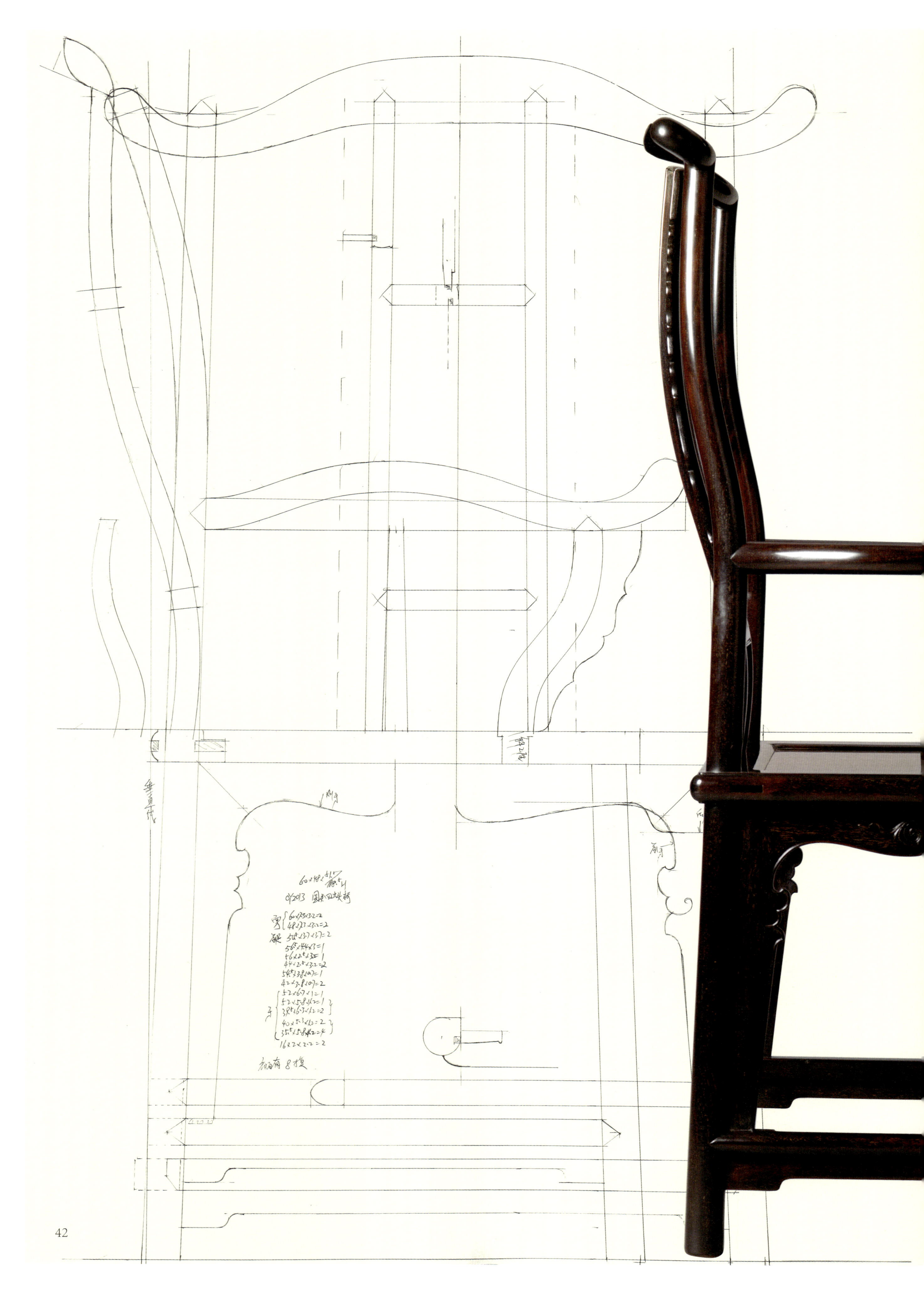

016

高靠背南官帽椅

High Yoke (Official's Hat)-back Armchair

59×47×51.5/119cm

椅子由黄花梨制成，选料精良，纹理流畅，色泽莹润。

搭脑曲线中部高起，向两侧顺势而下，至尾部作烟袋锅做法，与后腿上沿榫接。

搭脑之下装独板靠背板，靠背板修长高挑，精选纹理优美的木材，任其光素，尽显行云流水的自然流畅纹理，也是整件家具的点睛之处。扶手作弯曲曲线，收尾处做烟袋锅做法，与鹅脖榫接。扶手下，后腿与鹅脖之间装联帮棍，既增加坚固，又是造型细节处理。

座面四面攒边，面心装席面软屉，席面精心编织出精美纹饰。边抹做冰盘沿压边线。座面下接四腿，四腿出挓，看面两腿之间装素牙条券口牙子，任其光素。四腿之间装步步高赶枨，寓意步步登高。

整件椅子靠背较一般椅子高出一截，既凸显了独板靠背天然的优美纹理，又拉高了视觉中心，使得整件椅子颀长秀美，气质出众。

It's made of yellow *Huali (Huanghuali)* wood of refined material with smooth veining and elegant gloss.

The headrest slightly rises up in the middle at the joint of headrest and backrest and then runs lower on two sides, with two ends butted with rear legs by pipe-style joints.

Below the headrest installs an one-piece backrest of selected material with slender shape and beautiful natural veining as the visual focus of the furniture; armrests are processed into s-shape with two ends butted into front legs by pipe-style joints; below armrests, joint rods are installed to reinforce the steadiness.

Seat panel is constructed with assembled frames, a soft woven cushion and rattan surface of beautiful decorations; edges are narrowed down widthways (also called ice-plate edge) and carved with a rabbet along the bottom rim; below the seat installs four splayed legs; plain straight apron plates and stretchers are installed between legs to imply the wish of being promoted to a higher position.

The backrest of this chair is higher than the general ones to emphasize the gorgeous natural veining of the one-piece material and also raise the visual focus, giving a unique sense of slimness and elegance.

明万历
唐振吾刻《镌新编全相霞笺记》插图
广庆堂藏板

Wanli Period of Ming Dynasty (1573-1620)
Illustration of *Complete Story of Rosy Letters* (carved by Tang Zhenwu), Guangqingtang edition

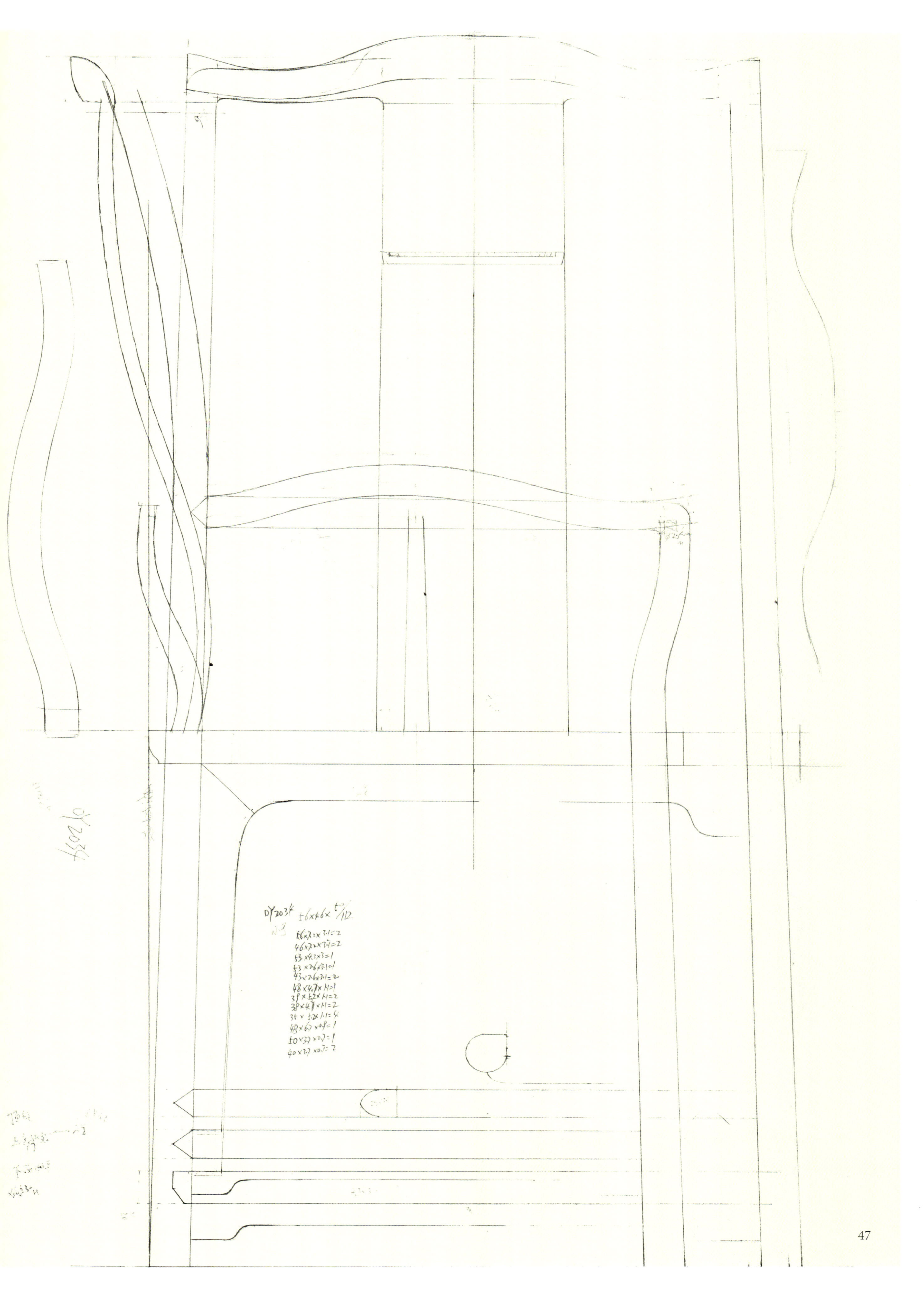

017

双螭纹南官帽椅

Yoke (Official's Hat)-back Armchair Embedded with Ivory Carving of Double-dragon Design

63×50×46/96cm

椅子由紫檀制成，色泽沉穆稳重。

搭脑做C形内收曲线，中间略粗，两端略细，收尾处做挖烟袋锅做法，与后腿榫接。相接处安云头角牙，勾勒空间曲线。搭脑下装S形独板靠背，靠背偏上部分镶嵌纹饰做开光，为双螭捧寿纹。

扶手S形，自起始到收尾顺势而下，做较大的倾斜，可支撑胳膊休憩。扶手与鹅脖亦为挖烟袋锅做法榫接。扶手下安S形联帮棍，曲线起伏较大，增加座内空间。

座面四面攒边，面心做席面软屉。边抹做泥鳅背，光素饱满。座面下安四腿，四腿出挓，以增加稳定性。两腿之间安素牙条券口牙子，边缘起阳线。四腿之间装赶枨。

此件椅子造型精炼，靠背板上的装饰是整件椅子的点睛之处，两侧角牙的修饰成为陪衬，相得益彰。

清康熙
梁延年编《圣谕像解》插图
承宣堂刊本

Kangxi Period of Qing Dynasty (1662-1722)
Illustration of *Illustrated Stories From Emperor's Edict* (edited by Liang Yannian), Chengxuantang edition

The ivory embedded chair is made of red sandalwood with dark and solemn colour and elegant gloss.

The headrest is processed into c-shape and slightly bent inward, broader in the middle area and gradually tapering toward two ends which are butted with rear legs by pipe-style joints and reinforced with two corner apron plates carved in cloud pattern; below the headrest installs an one-piece backrest in s-shape; a carving of double-dragon design is embedded at the upper part of the backrest, depicting an auspicious image of two dragons holding up the Chinese character of "寿" (longevity).

Armrests are processed into s-shape and tilted down heavily toward front legs to support people's arms, and butted with curved rods ("goose neck") by pipe-style joints; below armrests, it installs s-shaped joint rods with obvious undulation to add variation to spatial structure.

Seat panel is constructed with assembled frames and a soft woven cushion; edges are polished in loach-back style with fuller cambered appearance; below the seat installs four splayed legs to enhance the steadiness; straight apron plates outlined with embossed lines and lower stretchers are connected between legs.

This chair is of beauty of simplicity and quite distinguished in its embedded ivory carvings echoed by hollowed cloud-pattern corner apron plates.

018

夔龙纹三攒靠背南官帽椅

Yoke (Official's Hat)-back Armchair with Three-sectioned Assembled Backrest and Bas-relief of Dragon Pattern

58×47×50/98.5cm

椅子由黄花梨制成，黄花梨精选上等好料，纹理流畅优美，色泽温润，油性很大。

靠背三段攒成S形，上端在木板上浅浮雕夔龙纹饰，纹饰方形委角，周圈打洼。中部平镶纹理优美的黄花梨板，以纹理取胜。下端安壶门卷草纹亮脚，增加空间的通透性，边缘起阳线勾勒曲线。扶手后部高，末端矮，成明显的倾斜趋势，其功能与圈椅相类，可以承接胳膊。扶手后部接在后腿上沿，前部与鹅脖连接，烟袋锅做法，不设联帮棍，任其空间开敞。

座面四面攒边，面心装席面软屉。边抹泥鳅背，上下皆压边线。座面下接四腿，四腿出挓明显，形成下大上小的稳定感。两腿之间安素牙条券口牙子，边缘起阳线，勾勒曲线。四腿之间装赶枨，两侧略高，前面踏脚枨和后枨略低，这样做是为避开枨与腿之间的榫卯，增加稳定性。

此件椅子造型沉稳，扶手的做法不同于常规。此件椅子是复制王世襄《明式家具研究》中的甲75号椅子，不同之处在于原件靠背上部是镶嵌雕龙纹玉片，而此件改为黄花梨浮雕。

It's made of yellow *Huali (Huanghuali)* wood of highly refined material with smooth veining, elegant colour and rich oiliness.

The s-shaped backrest is assembled with three sections which respectively are carved with a bas-relief of dragon pattern framed with an embossed square with dented corners (upper), a yellow *Huali* plaque with beautiful veining (middle), and a corner plate carved in hollowed arabesque design outlined with embossed borderlines to add a sense of celestial beauty (bottom); armrests are tilted down toward front legs to hold arms, and butted into the upper part of rear legs and curved rods ("goose neck") by pipe-style joints without joint rods to leave more space open on the side.

Seat panel is constructed with assembled frames and a soft woven cushion with rattan surface; edges are polished in loach-back style with two rabbets along upper and bottom rims as borderlines; below the seat installs four splayed legs to reinforce its steadiness; straight apron plates outlined with embossed borderlines are installed between legs; stretchers are connected between legs with different heights: stretchers on the front and back are lower than those on two sides, so as to avoid the overlapping of multiple joints.

This chair is of solemn appearance with quite unique armrests distinct from common styles; it's of similar structure with chair A75 (甲 75) listed in *Connoisseurship of Chinese Furniture: Ming and Early Qing Dynasties* by Wang Shixiang, except that the original one mentioned in the book has a jade dragon carving embedded on the backrest instead of a bas-relief.

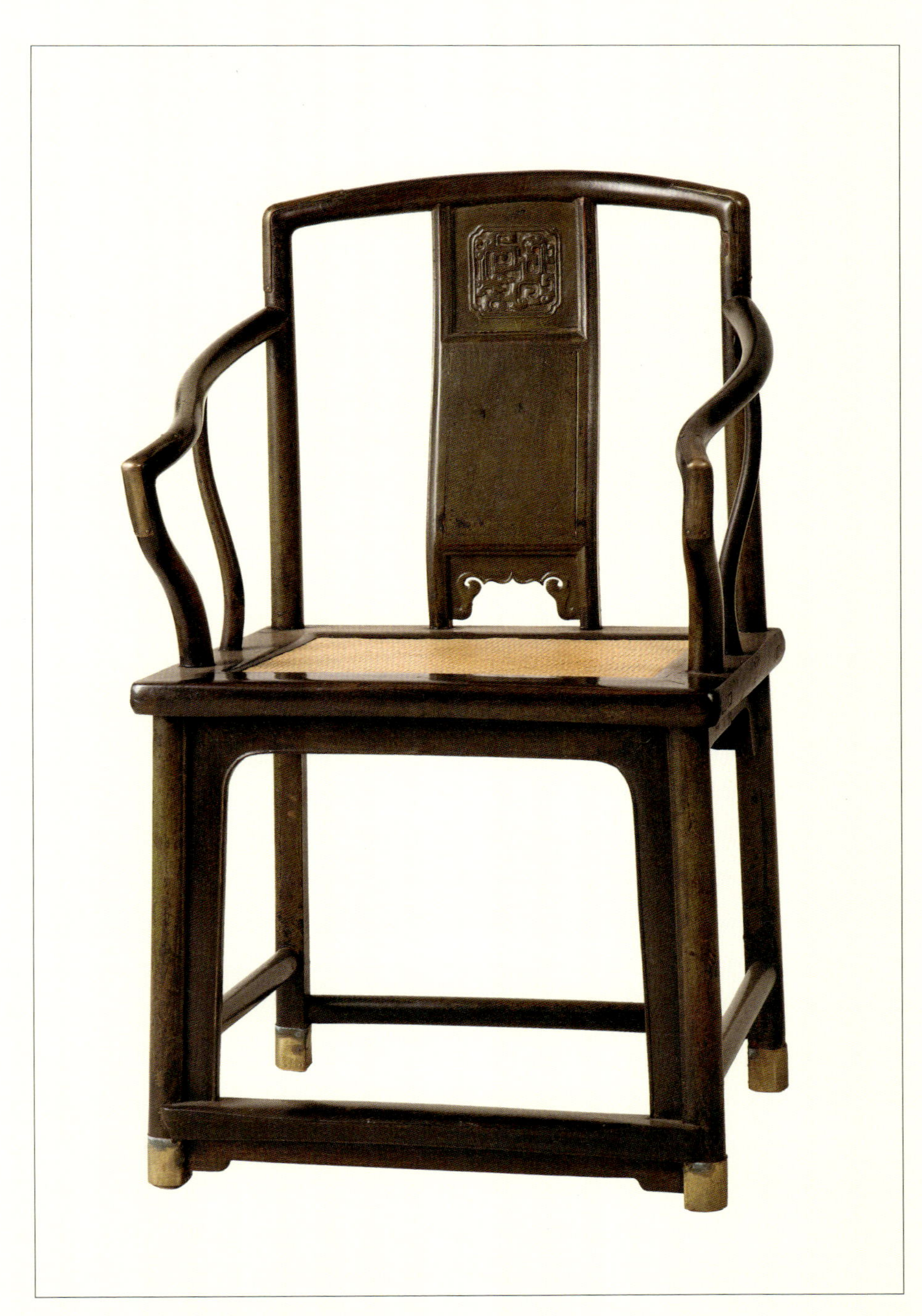

清　榉木夔龙纹三攒靠背南官帽椅（后刷漆）
区氏家具艺术馆藏品

Qing Dynasty (1644-1911)
Beech Yoke (Official's Hat)-back Armchair with Three-sectioned Assembled Backrest and Bas-relief of Dragon Pattern (lacquered)
From Family Ou's Furniture Collection

019

如意云纹三攒靠背南官帽椅

Yoke (Official's Hat)-back Armchair with Three-sectioned Assembled Backrest and Hollowed *Ruyi* Design

58×45×49/98cm

官帽椅由白酸枝镶嵌瘿木而成，质感温润，色泽淡雅。

搭脑弯曲甚多，兼以圆转的扶手，形成环身若抱的造型，倚坐甚为舒适。搭脑处的曲线中间粗而两头渐细，柔婉变化，外柔内刚，遒劲有力。靠背板三攒而成，沿边翻挺拔的细线，强调轮廓线。上段镶装绦环板，委角方形开光内升起如意云头，并有两个宝珠与两边连接，以求牢固，形成简洁大方的装饰效果；中段镶嵌瘿木，为“满地葡萄”纹，灿若火焰；下端镶壸门亮脚牙板，曲线大方，与上段的如意云纹遥相呼应，取得了空灵的装饰效果。

座面软屉，冰盘沿下方压边线。正面装壸门券口牙板，曲线极佳，内蕴力量，外显柔和。侧面装垂洼膛肚券口牙板，背后则装刀牙板。腿足外圆内方，设步步高赶枨，踏枨下设牙条。

此椅整体造型落落大方，装饰得体，饱满圆润，气势俨然，令人见之难忘。

It's made of white dalbergia cochinchinensis and embedded with a burl plaque in the backrest, with a mild jade-like texture and elegant colour.

The headrest is elegantly curved, with two undulating armrests, gently embracing the seat panel; headrest is thicker in the middle and gradually tapers toward two ends to form a steady inner structure and graceful appearance; backrest is assembled with three sections outlined by embossed borderlines, and respectively embedded with a decorative panel carved out a hollowed *Ruyi*-shaped cloud pattern framed by a square tracery with dented corners and fixed by two pearl-shape chips (upper), a burl plaque of grape vine veining (middle), and a hollowed *Kun*-gate corner apron plate (bottom) to echo the carving of cloud pattern on the upper side, giving a sense of celestial beauty.

Seat panel has a soft woven cushion and edges narrowed down widthways (also called ice-plate edge), with a rabbet along the bottom rim; the *Kun*-gate apron plates between front legs are of beautiful curves, exhibiting strength and sexual tenderness; between legs on two sides installs straight apron plates, and on the back installs an apron plate by T-bridle joints (clamp joint); legs are of half-round-and-half-square sections and butted with stretchers on lower side; and an extra straight apron plate is installed below the footrest.

This chair is of decent structure and properly adorned, exhibiting an impressive sense of solemnity.

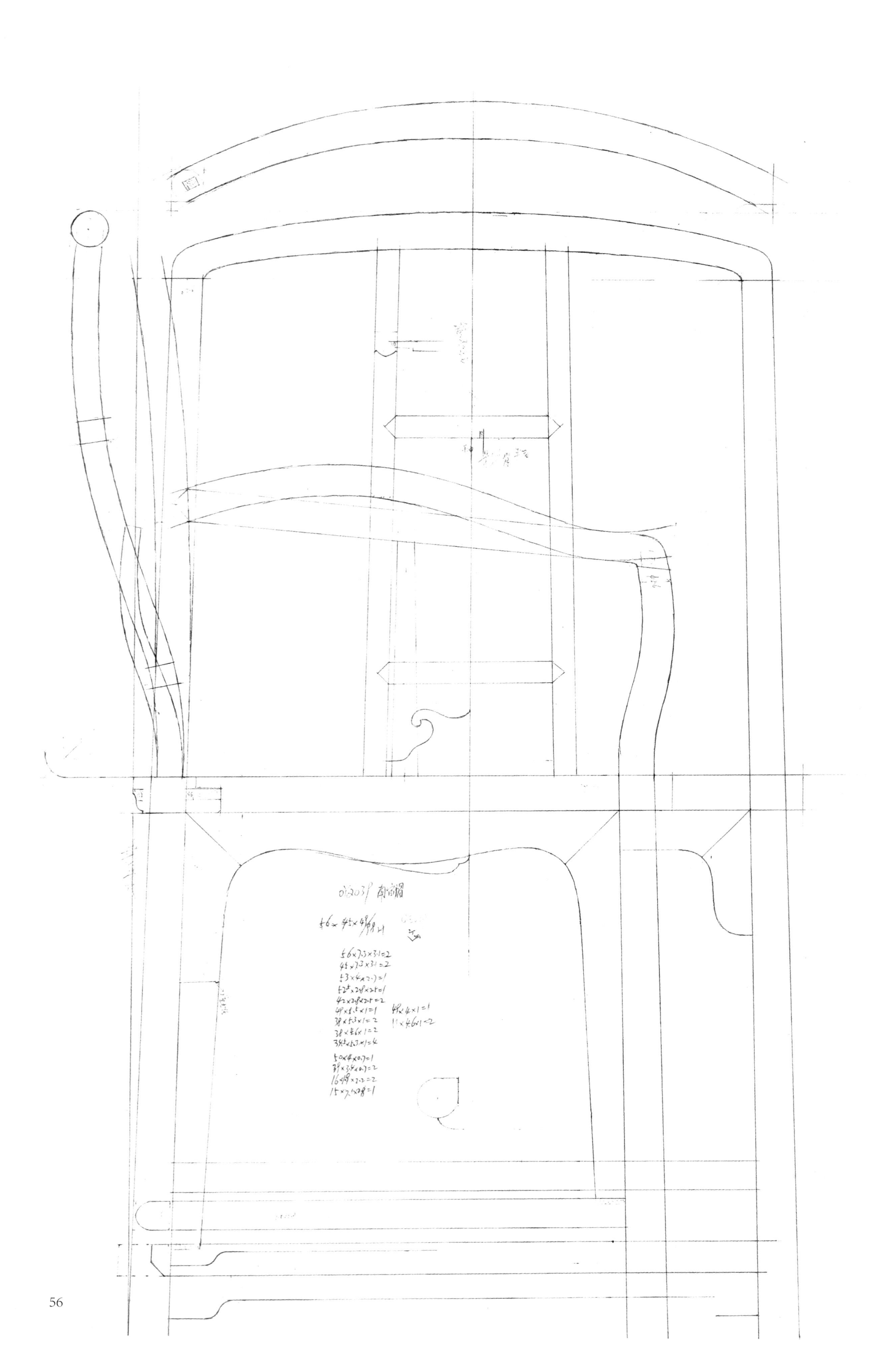
012031 南官帽
56×7.3×3.1=2
45×7.3×3.1=2
53×4×2.7=1
52.5×2.8×2.5=1
42×2.8×2.5=2
49×5.5×1=1
38×5.3×1=2
38×5.6×1=2
34.5×5.3×1=4
50×4×0.7=1
39×3.4×0.7=2
16×49×2.2=2
49×4×1=1
11×4.6×1=2

佚名《燕寝怡情》册页　美国波士顿美术馆藏

Illustration of *Noblemen's Everyday Life* (album)
by anonymity
From Museum of Fine Arts, Boston

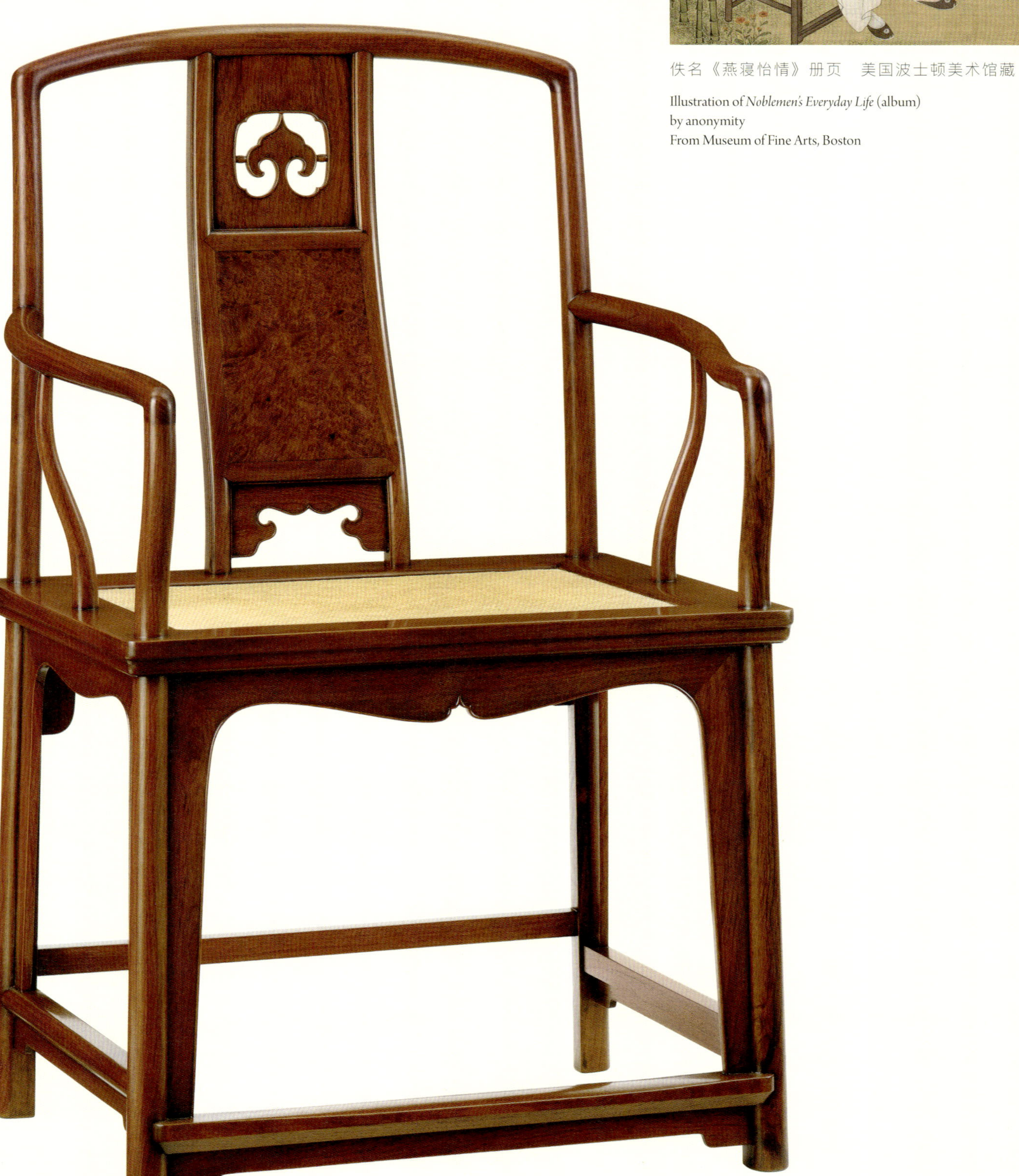

020

圈椅

Horseshoe-back Armchairs (a Pair)

60×46.5×51.7/102.7cm

此椅成对，由黄花梨制成，黄花梨纹理优美，色泽莹润，油性充足。

椅圈五截圈，以楔钉榫榫接成饱满流畅的圆形椅圈，椅圈于末端收尾处自然外展，以鳝鱼头形收尾。椅圈正中间连接独板靠背，并素面示人，以展现黄花梨优美舒畅的自然纹理。圈椅四腿接穿过座面向上，与椅圈直接连接。前腿上沿和后腿上沿之间装S形联帮棍，曲线舒展外延，增加了座面上的空间。

座面四面攒边，面心装席面软屉。边抹做冰盘沿压边线。座面下四腿出挓，四腿截面内方外圆。两腿之间装素牙条券口牙子，边缘起阳线勾勒轮廓。四腿下装步步高赶枨。

此圈椅刻意不着修饰，以尺度胜，以造型胜，以自然纹理胜，精心设计，又无刻意痕迹，此乃高明之处。

此对圈椅成对制作，难得之处在于下料皆精挑细选，两块独板靠背也是一块黄花梨木料下料得来，故而纹理相似又略有变化。

This pair of armchairs are made of yellow *Huali (Huanghuali)* wood of refined material with gorgeous veining, elegant gloss and rich oiliness.

The horseshoe-shaped headrest consists of five components fixed by dowels, with two ends slightly bent outward and polished in eel-head style; the one-piece backrest is installed to the middle part of headrest, and left unadorned to better demonstrate the gorgeous natural veining of yellow *Huali* wood; four legs are directly butted into headrest through seat panel; s-shaped joint rods are installed between upper ends of legs above the seat, with graceful curve undulating outward to expand the space above.

Seat panel is constructed with assembled frames and a soft woven cushion of rattan surface; edges are narrowed down widthways (also called ice-plate edge) with a rabbet along the bottom rim as borderline; below the seat installs four splayed legs with half-round-and-half-square sections; straight apron plates outlined by embossed lines and lower stretchers are installed between legs.

This unadorned armchair epitomizes the aesthetics of simplicity, is quite distinguished with its form and scale, as well as the natural veining and texture of the material, presenting a rarely exquisite yet subtle beauty.

This two armchairs are produced in pair with highly selected material; two backrests are also carved out a same yellow *Huali* material, hence the similar veining.

021

团螭纹靠背圈椅

Horseshoe-back Armchairs Carved with Dragon Design (a Pair)

60×46.5×51.5/101.5cm

此圈椅成对，由花梨木制成，选材精良，纹理优美，色泽温润。

椅圈为五截圈，椅圈曲线饱满舒畅，于收尾处外展，收为鳝鱼头形。椅圈正中装C形靠背，靠背由两侧的竖枨装独板，竖枨做剑脊棱线脚，独板偏上开如意云头形开光，镶嵌黄杨透雕的团螭纹，下端开云头纹亮脚。以竖枨装板的做法，既可以给人独板靠背的感觉，心板又可以用薄板制作以省料。前腿和后腿皆穿过座面向上与椅圈榫接，后腿上沿两侧装长条角牙。

座面四面攒边，面心装席面软屉，边抹做冰盘沿压边线。座面下四腿出挓，四腿截面内方外圆。四腿之间装壶门券口牙子，边缘起阳线以勾勒轮廓。四腿之间偏下装步步高赶枨。

此圈椅尺度合宜，造型利落，曲线饱满，气宇不凡。

This pair of armchairs are made of *Huali* wood of refined material with graceful veining and mild gloss.

The horseshoe-shaped headrest consists of five components fixed by dowels, with two ends slightly bent outward and polished in eel-head style; an one-piece c-shaped backrest is embedded between two vertical stretchers with sword-ridge rabbets and butted into the middle area of headrest; the backrest plaque has a hollowed boxwood carving of dragon outlined with embossed borderline of cloud-pattern embedded on upper side, and has the bottom carved in hollowed cloud-pattern; such backrest structure can stress the integrality of the one-piece plaque and as well save the material by using thin plate; four legs are directly butted into headrest by dowels through seat panel; straight vertical apron plates are installed to rear legs on two sides above the seat.

Seat panel is constructed with assembled frames and a soft woven cushion; edges are narrowed down widthways (also called ice-plate edge) with a rabbet along the bottom rim as borderline; below the seat installs four splayed legs with half-round-and-half-square sections; *Kun*-gate apron plates outlined with embossed borderlines and lower stretchers are installed between four legs.

This pair of armchairs are of proper scale and clean structure, presenting an outstanding and solemn beauty.

清　佚名《隐睿西厢记》册
故宫博物院藏

Qing Dynasty (1644-1911)
Illustration of *Romance of the Western Chamber* (album)
by anonymity
From Palace Museum, Beijing

022

螭龙纹靠背圈椅

Horseshoe-back Armchair Carved with Dragon Design

60×46.5×51.7/101.7cm

圈椅由花梨木制成，精选好料，巧借木材天然的纹理，木料油性充足，色泽莹润。

椅圈为五截圈，以楔钉榫榫接，椅圈曲线圆润饱满，于近扶手处自然外展，以鳝鱼头形收尾。椅圈正中连接独板宽靠背，靠背纹理精选，流畅自然，靠背偏上浅浮雕螭龙纹开光。靠背上小下大，成稳定之势。靠背与椅圈连接处做长条角牙，角牙与靠背板一木连做。四腿皆穿过座面向上，与椅圈连接。后腿上沿两侧装长条牙条，而鹅脖与椅圈连接处装角牙，皆为增加连接坚固的作用。后腿上沿和前腿上沿之间装联帮棍，S形曲线。

座面四面攒边，面心席面软屉。边抹做混面压边线。座面下四腿出挓，两腿之间装壶门牙条券口牙子，壶门牙条之上浅浮雕卷草纹饰，两侧牙子翻卷出卷草。更为独特之处在于，券口牙子上部亮脚，与座面和腿交界处，浅浮雕出似铜包角的云头曲线，较为少见。四腿之间偏下装步步高赶枨。

此圈椅在比例、尺度上考量有度，加之靠背和券口牙子适当的装饰，为此件圈椅增添了雍容典雅的气质。

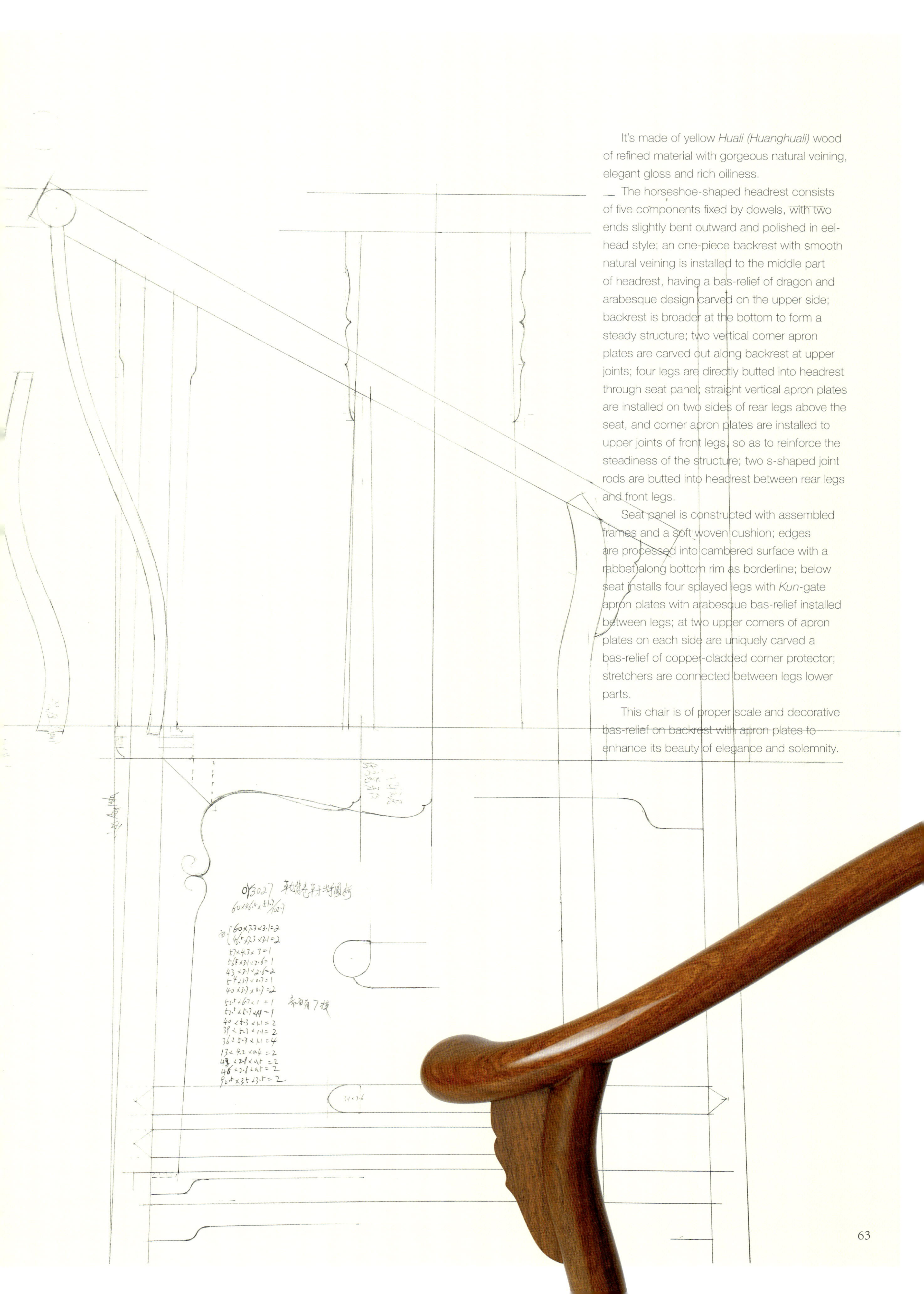

It's made of yellow *Huali (Huanghuali)* wood of refined material with gorgeous natural veining, elegant gloss and rich oiliness.

The horseshoe-shaped headrest consists of five components fixed by dowels, with two ends slightly bent outward and polished in eel-head style; an one-piece backrest with smooth natural veining is installed to the middle part of headrest, having a bas-relief of dragon and arabesque design carved on the upper side; backrest is broader at the bottom to form a steady structure; two vertical corner apron plates are carved out along backrest at upper joints; four legs are directly butted into headrest through seat panel; straight vertical apron plates are installed on two sides of rear legs above the seat, and corner apron plates are installed to upper joints of front legs, so as to reinforce the steadiness of the structure; two s-shaped joint rods are butted into headrest between rear legs and front legs.

Seat panel is constructed with assembled frames and a soft woven cushion; edges are processed into cambered surface with a rabbet along bottom rim as borderline; below seat installs four splayed legs with *Kun*-gate apron plates with arabesque bas-relief installed between legs; at two upper corners of apron plates on each side are uniquely carved a bas-relief of copper-cladded corner protector; stretchers are connected between legs lower parts.

This chair is of proper scale and decorative bas-relief on backrest with apron plates to enhance its beauty of elegance and solemnity.

023

太狮少狮纹三攒靠背圈椅（附方几）

Horseshoe-back Armchair with Three-sectioned Assembled Backrest Carved with a Bas-relief of a-lion-playing-with-his-cub (with Square End Table)

61.5×47.5×50/100cm

圈椅由白酸枝制成。五截圈，围合成近圆的曲线，于近扶手处内弯再外展，成鳝鱼头式收尾。椅圈正中连接靠背，靠背三段攒成，上段装板透雕云头纹；中段最长，装板浅浮雕太狮少狮纹，狮子活跃，长尾巴翘起，作回首状，一爪按住下方小狮子，生动传神；下端作云头纹开光。

四腿皆穿过座面向上延伸，与椅圈榫接。后腿上沿两侧装长条牙板，前腿上沿与扶手连接处亦装长牙条。S形联帮棍连接扶手和座面，勾勒侧面空间。

座面四面攒边，面心装硬屉，边抹做冰盘沿压边线。座面下接四腿，腿间装壸门牙条券口牙子，壸门牙条之上浅浮雕卷草纹，草叶翻转舒展自如，两侧牙条偏上内翻草叶，与壸门曲线相呼应。边缘起阳线以勾勒轮廓。四腿之间偏下装步步高赶枨。

整件圈椅在合宜的尺度、比例基础之上做适当装饰，特别是靠背板的装饰处理，是本件圈椅的出彩之处。

几方形，几面四面攒边，格角榫连接之处，大边出小舌与抹头相接，是增加坚固的细节处理，结构到位。面心平装硬板。

几面之下接四腿，腿截面内方外圆，四腿八挓，造型上敛下舒有稳定感。两腿间座面下安花牙子，牙条平直，牙头做卷草。花牙子边缘起阳线勾勒轮廓。花牙子之下紧挨着装罗锅枨，罗锅枨曲线略起，刚劲有力。四腿修长落地，偏下之处周圈装横枨和素牙子。

此件方几造型简洁洗练，比例修长优雅，着墨少许做装饰，恰到好处。与圈椅、官帽椅皆可完美搭配。

It's made of white dalbergia cochinchinensis. The horseshoe-shaped headrest consists of five components fixed by dowels, with two ends slightly bent outward and polished in eel-head style; a three-sectioned assembled backrest is installed to the middle part of headrest, respectively embedded with a hollowed carving of *Ruyi*-shaped cloud pattern (upper), a plaque carved with bas-relief of a-lion-playing-with-his-cub (middle), and a hollowed corner plate carved with cloud patterns (bottom); the carving on middle plaque depicts a lion springing down and raising his tail, caressing his cub with right paw.

Four legs are directly butted into headrest through seat panel; vertical straight apron plates are installed on two sides of rear and front legs above the seat; s-shaped joint rods are butted into headrest to outline the side silhouette.

Seat panel is constructed with assembled frames and a hard cushion, with edges narrowed down widthways (also called ice-plate edge) and carved a rabbet along bottom rim as the borderline; below seat installs four legs, having *Kun*-gate apron plates with intertwined arabesque bas-relief and straight plates with floral carvings on upper sides installed between legs; all the plates are outlined with embossed borderlines; lower stretchers are connected between legs.

This chair is of proper scale and highlighted by decorative bas-relief on backrest.

Seat panel of the square end table is constructed with a hard cushion and assembled frames butted by mitered corner bridle joints with through splines to reinforce the steadiness.

Below the seat installs four splayed legs with half-round-and-half-square sections to give a sense of stability; floral apron plates outlined with embossed borderlines are butted between legs below the seat; hunched rails are installed between legs below apron plates to exhibit strength and sturdiness; stretchers butted with apron plates are installed at the lower part of four legs.

This square end table is of simple and clean form, elegant and slender appearance, and proper decorative bas-relief, as a perfect match for other armchairs.

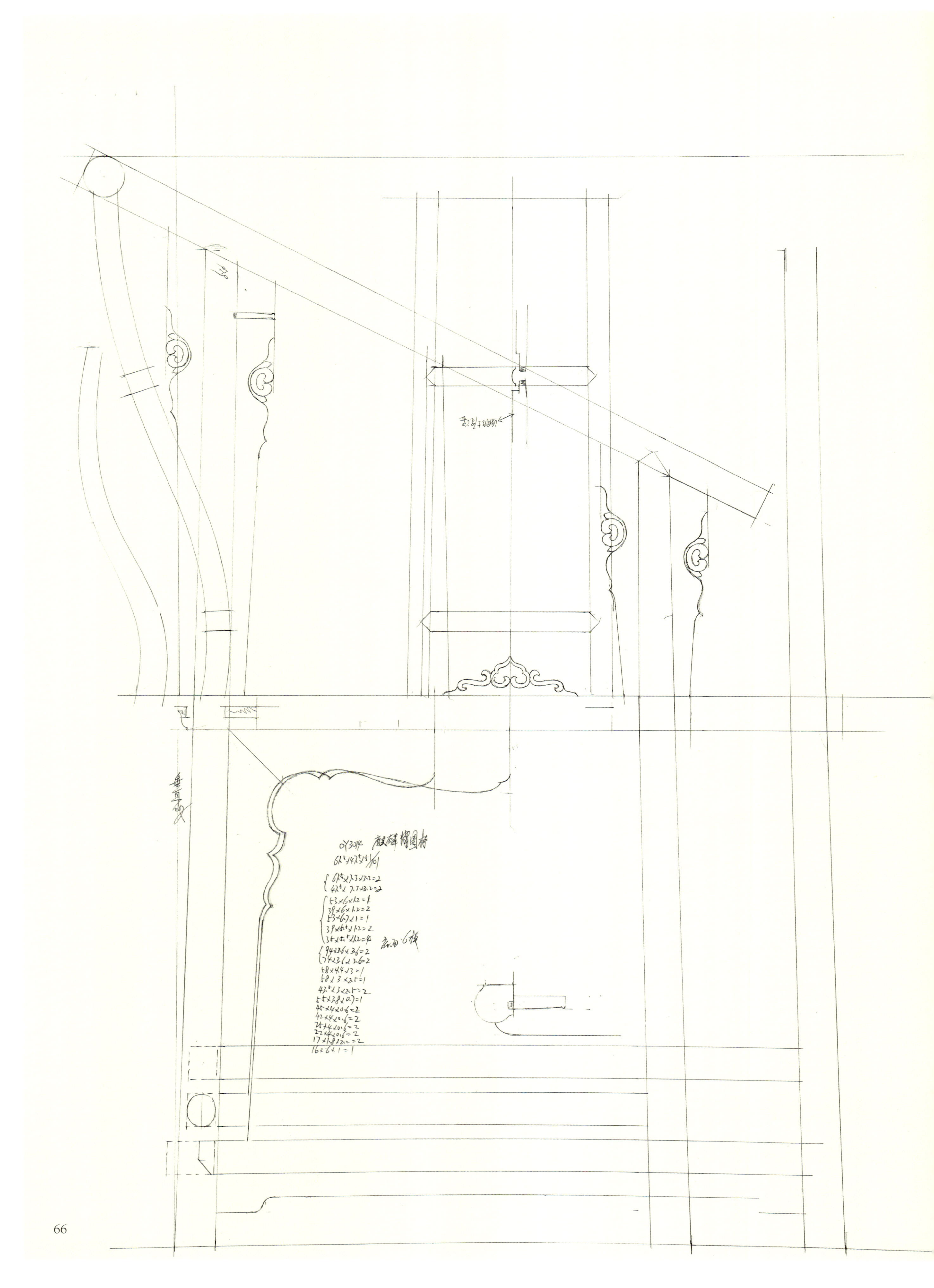

024

镶大理石如意形开光三攒靠背圈椅

Horseshoe-back Armchair with
a *Ruyi*-shaped Tracery

53×41×46.5/91.7cm

圈椅白酸枝镶大理石而成。其扶手不出头，是圈椅中较少见的品种，也正因此，其椅圈扶手顺畅自然地过渡至鹅脖，流畅无碍，比出头圈椅更加柔婉含蓄，所见这类圈椅，几乎无不精彩。

椅圈三接，鹅脖较高，椅圈平缓，兼以素直的鹅脖，形成甚为文气的造型。靠背板三攒式，横竖枨边沿皆起边线强调，上段设如意形开光，通透敞亮；中段圆光内镶大理石一块，若月之东升，醒目而气象万千；下段设壶门亮脚牙板，曲线柔婉而窄小，所余空间较大，与整体空灵的意趣相合。设S形三弯联帮棍，与靠背板曲线呼应，袅袅而起，若风动春草，甚具韵律感。

座面攒框装软屉，边缘素混面，转折处有自然的棱线。腿间装前后低赶枨，四面装直素的券口牙板，沿边起碗口线，保持整洁素雅的风格。

原物为原美国加州古典家具博物馆所藏，一套四件，镶嵌大理石各不相同，宛然成春夏秋冬之景，是圈椅中的经典代表。

This marble-embedded chair is made of white dalbergia cochinchinensis; the horseshoe-armrest is uniquely butted with straight rods (also called "goose neck") by pipe-style joints as a rare case in armchairs, which outlines a smooth silhouette and exhibits a more feminine and subtle aesthetics.

The horseshoe-headrest consists of three components fixed by dowels, with its two ends slightly tilted down toward front legs and butted with straight rods ("goose neck") which are higher than usual by pipe-style joints to demonstrate an aesthetics of literati; backrest is assembled with three sections respectively embedded with a *Ruyi*-shaped tracery (upper), a round marble plaque with natural veining resembling a rising moon hanging above a mountain (middle), and an exquisite *Kun*-gate apron plate, leaving much space open to maintain a unified ethereal beauty (bottom); s-shaped three-curved joint rods are installed to echo the curvature of backrest, showcasing a rhythmic beauty.

Seat panel is constructed with assembled frames and a soft woven cushion, with unadorned cambered edge; lower stretchers and plain straight apron plates outlined with embossed borderlines are installed between legs to maintain a unified clean and simple style.

The original one is collected by the museum of classical Chinese furniture (California, USA), among a set of four embedded with various marble plaques seemingly representing four seasons. This set of armchairs are quite typical and of high aesthetic value.

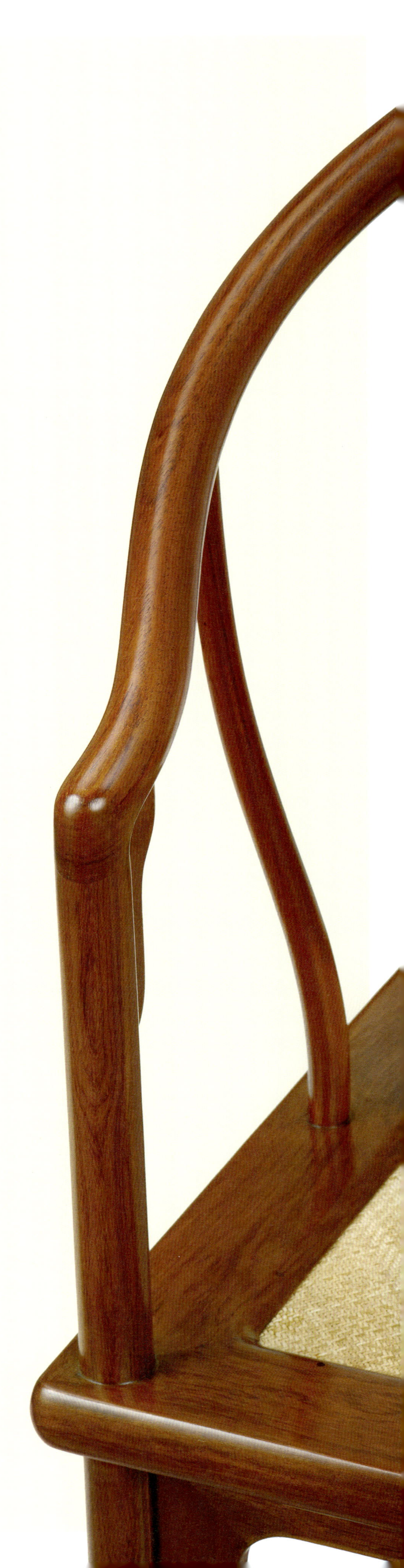

025

仿藤式圈椅

Horseshoe-back Armchair in Vine-style

60×46.5×48.5/98.5cm

圈椅以铁梨木制成。前已述及，不出头圈椅为圈椅中柔婉的一类，此件造型则是将柔婉的造型表达至极致，究其根本，则是因造型借鉴了藤质家具的特征，交接处处柔和，内蕴力量。

圈椅扶手内收后又撇出，鹅脖上部则后弯后与之相接，形成蜿蜒的效果。联帮棍、靠背板的三弯造型亦较夸张，与整体呼应。座屉边框素混面，其下牙板为三段扁圆枨攒成，转角圆转相接，中段垂洼膛肚，如弯藤而成，甚为别致。券口牙板如此做法，为目前仅见的孤例。圆腿足，装步步高赶枨，前、左、右三面枨下设牙条加固。

圈椅取藤条弯折圆转之意趣，将坚硬的铁梨木加工成柔婉的造型，处处用心，在一众圈椅中脱颖而出，令人过目难忘。更需提出的是，因尺寸大，比例开张，虽然模拟藤质柔婉圆转的造型，却未失于矫揉造作，显示出制作者高超的把控能力。

原作为著名美国收藏家安思远旧藏。

It's made of lignumvitae. This armchair is the superlative expression of softness as its structure and appearance of components imitate vines; the horseshoe-armrest is uniquely butted with curved rods (also called "goose neck") by pipe-style joints to outline a smooth silhouette and exhibits a more feminine and subtle aesthetics.

Armrests gradually converge toward front legs and are slightly bent outward at end; upper ends of front legs are bent backward and butted with two ends of armrests to continue the undulating outline; joint rods and backrest are also processed into obvious three-curved shape to echo the general style; below the cambered edge of the seat panel installs three sections of cylindrical rails butted into legs and seat edge with rounded corners; the middle part of the horizontal cylindrical rails slightly drops to resemble the appearance of vines; it is a rare and only case known to us so far that cylindrical rails are applied as apron plates; legs with round sections are connected with stretchers fixed with apron plates below.

This chair transforms the hard and rigid lignumvitae to a representation of softness and pliableness through its resemblance to vines in form and structure, absolutely outstands the common kinds. Moreover, due to its scale, though it imitates turns and curves of vines, it still gives a sense of solemnity, as a demonstration of craftsmen's control over form and structure.

The original one was from Robert Hatfield Ellsworth's former collection.

026

螭龙捧寿纹玫瑰椅（附方几）

Low-back Armchair (also called Rose Armchair) with Tracery and Bas-relief of Dragon Design

61×46×48/88cm

玫瑰椅是一种靠背较矮，接近扶手，且多为直型的椅具。其三面装饰，简洁者装栏杆状构件、券口牙板，复杂者镶透雕花板，富丽堂皇。此例属后者。

玫瑰椅以花梨木制成，搭脑和扶手下三面靠下装横枨，下设团螭纹卡子花，靠背处横枨上方，镶装花板，透雕兼浮雕螭龙捧寿纹，中间如意形图案由螭龙变体而成，围绕着一个同样是螭龙纹变体而成的鼎形寿字，周围布满大小不一、简繁不同的螭龙纹，姿态万千，繁而不乱。两侧扶手下设壶门券口牙板，亦浮雕螭纹。

座面攒框落膛装板，四周设冰盘沿。腿足外圆内方，之间设步步高赶枨，四面设券口牙板，沿边起线，衍为回纹。上方牙条处亦浮雕螭龙纹，身躯若兽，随牙板曲线而起伏，甚具动感。

附有束腰带管脚枨方几，牙板坐壶门状，与玫瑰椅呼应。

此椅原物两对，一对为故宫博物院所藏，一对藏清华大学艺术博物馆，曾收入王世襄《明式家具研究》及《明式家具珍赏》，是玫瑰椅之经典样式。

清　郑崑屿《仕女坐像》
加拿大安大略皇家博物馆藏

Qing Dynasty (1644-1911)
Sitting Portrait of a Lady by Zheng Kunyu
From Royal Museum of Ontario, Canada

Low-back armchair (also called rose armchair) is with a lower backrest and straight outline, usually decorated on three sides (back and two sides): either simply adorned with components of handrails and apron plates, or plaques exquisitely carved with hollowed designs like this one.

It's made of yellow *Huali (Huanghuali)* wood. Below headrest and armrests installs three horizontal rails butted with delicate round chips carved with hollowed dragon patterns; a decorative plaque with openwork of dragons of different variations and postures holding up "寿" (Chinese character meaning longevity) is embedded in backrest above the horizontal rail; below the armrests installs *Kun*-gate apron plates carved with bas-relief of dragons.

Seat panel is constructed with assembled frames and planed down central plaque, with edges narrowed down widthways (also called ice-plate edge); legs with half-round-and-half-square sections are connected with stretchers and apron plates carved with bas-relief of crouching dragons and outlined with embossed borderlines and fretwork patterns.

The armchair comes with a square end table with girdled waist and foot rails, as well as decorative Kun-gate apron plates to coincide with the design of the armchair.

The original chairs come in two pairs: one pair are collected in Palace Museum (Beijing), the other in Tsinghua University Art Museum, listed in *Connoisseurship of Chinese Furniture: Ming and Early Qing Dynasties* and *Classic Chinese Furniture: Ming and Early Qing Dynasties* by Wang Shixiang, as the representative of classic low-back armchair.

027

券口牙板玫瑰椅

Low-back Armchair (also called Rose Armchair) with Apron Plates Butted Backrest

58×45.5×48/87cm

玫瑰椅以白酸枝制成，属于同类中较为简洁的造型，空灵舒朗，以结构之美打动人心。

直搭脑、扶手，三面靠下方设横枨、矮老，状如栏杆，靠背横枨上设券口牙板，沿边起线，并衍化为对卷勾云纹，其他一任空敞。座面攒框，装软屉，边框做成简素的冰盘沿样式，下方仅压边线。腿足外圆内方，三面设直券口牙板，继续保持简素空灵的造型风格，腿足下方装步步高赶枨，前、左、右下三面设牙条。

玫瑰椅因靠背较矮，搁置居室，既便使用，又不会遮挡视线，影响空间，因而是点缀空间的经典用具。

宋（传）李公麟《西园雅集图》 台北故宫博物院藏

Song Dynasty (960-1279)
West Garden Gathering
Allegedly by Li Gonglin
From Taipei Palace Museum

It's made of white dalbergia cochinchinensis, with clean and simple structure, giving a sense of ethereal beauty.

Below the straight headrest and armrests installs horizontal rails and rods in baluster style; apron plates are installed within the frame of backrest above the horizontal rail, and outlined with embossed borderlines and fretwork pattern, leaving the rest space open; seat panel is constructed with assembled frames, a soft woven cushion and edges narrowed down widthways (also called ice-plate edge) with a rabbet along the bottom rim as borderline; legs are of half-round-and-half-square sections with straight apron plates to maintain a simple and celestial beauty; stretchers are installed at lower part of four legs; and straight apron plates are butted below stretchers on front, left and right sides.

Due to its low backrest, this armchair can be placed in rooms without blocking sight or interfering the spatial fluency, hence a classical furniture to embellish the living space.

028

螭龙捧寿纹圈背交椅

Horseshoe-back Folding Chair with Bas-relief of Dragons Holding up Chinese Character "寿" (Longevity)

58×43×53/103cm

交椅由黄花梨制成。椅圈五截，衔接成圆润的曲线，自搭脑中部向两侧扶手顺势而下，弧度流畅自如，近扶手处内转再外翻出头，成圆润扶手。椅圈接榫处皆用铜片包裹，主要起加固作用，铜片边缘做云头曲线，又是一种细节的装饰。

靠背由一块独板雕刻为三攒式。上段点浮雕卷草纹饰，中段较长，浅浮雕双龙捧寿纹，下段浅浮雕云头曲线做亮脚。靠背与椅圈相交处装曲边角牙，起增加坚固作用。整个靠背皆用浮雕的处理方式，皆不透雕，通常做透光的亮脚，也只是曲线上示意而已，并不挖透。

前腿穿过椅面做大弧度转角，与椅圈接于近扶手处。又在前腿上沿和扶手连接处装雕花角牙，两者之间又安铜制立柱，以增加支撑。

椅面前大边浅浮雕草龙纹饰，与靠背雕琢相呼应。两大边之间装细木条为屉，可以折叠。座面之间为侧面两腿的转轴，可以开合。四腿之下承接托泥和榻床。四腿和托泥接榫处亦用铜片加固，榻床之上也包铜片一圈，并作适当纹饰装饰。

此件圆背交椅比例协调，适当雕琢装饰，巧用铜饰件，气质秀雅。

It's made of yellow *Huali (Huanghuali)* wood. The horseshoe-shaped headrest consists of five components fixed by dowels; it slightly rises up in the middle, then runs lower on two sides, and is finally bent upward on two polished ends. The butted joints of the headrest are wrapped by copper sheets outlined with cloud patterns to reinforce the curve structure.

The backrest carved out of an one-piece material is assembled with three sections respectively carved with bas-relief of floral pattern (upper), double dragons holding up Chinese character "寿" (meaning longevity) (middle), and cloud patterns as apron plate other than the usual openwork style (bottom); at joints of backrest and headrest installs two vertical corner apron plates with curved borderlines for reinforcement; the whole backrest plaque is decorated with bas-relief without any openwork.

Front legs are curved and directly butted with armrests through seat panel to reinforce the structure; corner apron plates carved with floral patterns and copper rods are butted into the joints of front legs and armrests to enhance the steadiness.

Seat panel's front edge is carved with bas-relief of dragons and floral patterns to echo the backrest; battens are laid across the seat panel as folding cushion; pivots are installed in the middle of legs on two sides; below four legs connects foot rails and a foot pad with joints reinforced by copper sheets and edges wrapped with copper sheets decorated with geometrical patterns.

This horseshoe-back folding chair is of proper scale, elegant decorative bas-relief and copper fittings to exhibit a graceful beauty.

明万历
《管鲍分金记》插图

Wanli Period of Ming Dynasty (1573-1620)
Illustration of *Story of Guan Zhong and Bao Shuya Dividing Earnings*

明万历
《御世仁风》插图
凤阳刊本

Wanli Period of Ming Dynasty (1573-1620)
Illustration of *Collected Stories of Emperors*,
Fengyang edition

029

螭龙捧寿纹圈背交椅

Horseshoe-back Folding Chair with Bas-relief of Dragons Holding up Chinese Character "寿" (Longevity)

58×43×53/103cm

交椅由紫檀制成，紫檀沉穆稳重，配以铜饰件，材质对比鲜明而又统一和谐。

椅圈五截，榫接处用云头曲线的铜片加固，椅圈又与靠背、后腿上沿和后腿连接，连接之处接嵌铜加固，使得整圈铜饰件满饰。

靠背由一块独板雕刻为三攒式。分别浅浮雕卷草纹、双龙捧寿纹和云头纹饰，成一整体。后腿穿过座面向上延伸，作S形曲线，与椅圈榫接，又配以角牙和铜立柱做加固。

座面大边看面浅浮雕草龙纹饰，两边挺之间安细木条串成的软屉，方便折叠收纳。座面下四腿交叉，用于折叠，转轴之处安铜制护眼钱以加固。四腿之下皆托泥，前腿上安榻床。

此件圆背交椅巧用不同材料的对比与呼应，又加以适当的装饰，使得整体协调一致，气质独具。

It's made of red sandalwood with dark and solemn colour, exhibiting a sharp contrast with copper fittings. The clash of different materials and textures creates a rare visual impact, while decorative designs are kept in a unified style.

The horseshoe-shaped headrest consists of five components fixed by dowels; and the butted joints of the headrest are wrapped by copper sheets to reinforce the curve structure; backrest, rear legs' shoulders and ends are directly butted with the headrest and reinforced by copper fittings.

The backrest carved out of an one-piece material is assembled with three sections respectively carved with bas-relief of floral patterns (upper), double dragons holding up Chinese character "寿" (meaning longevity) (middle), and cloud patterns; s-shaped rear legs are directly butted with two ends of headrest and reinforced with copper rods at legs' shoulders and corner apron plates at joints.

Seat panel's front edge is carved with bas-relief of dragons and floral patterns; battens are laid across the seat panel as folding cushion; below the seat, legs are crossed, and pivots wrapped with copper caps are installed in the middle of legs on two sides; below four legs connects foot rails and a foot pad on the front.

This horseshoe-back folding chair showcases a dramatic contrast between different materials, yet still keeps a unique harmonious uniformity by proper decorations.

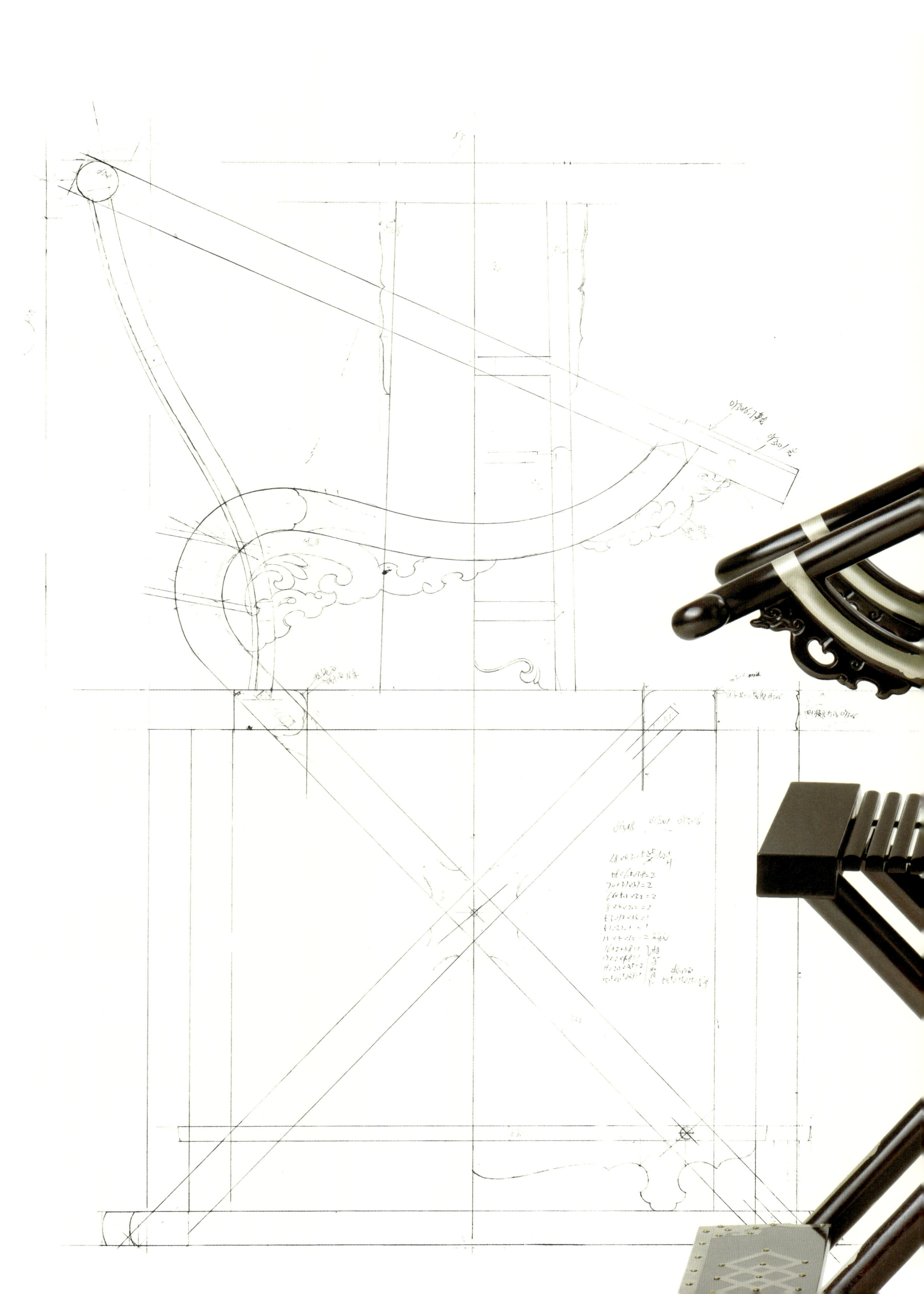

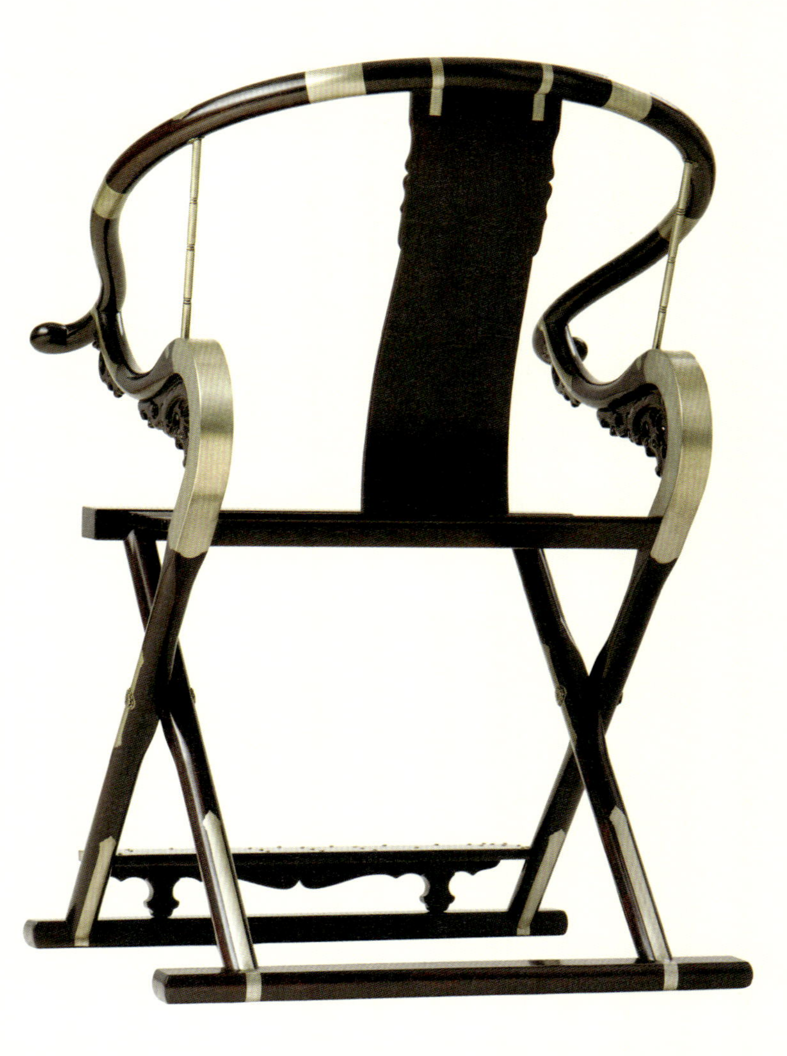

030

镶大理石交椅式躺椅

Folding Lounge Chair Embedded with Marbles

59×49×44/96cm

躺椅由白酸枝制成，选料精良，纹理优美，色泽莹润，编织席面靠背，两种材料质感、肌理和色彩不同中呈现统一。

交椅式躺椅结构以转轴为中心，进行造型处理。前腿经由转轴至后方，向上延伸为两侧框架，框架之间装横枨成靠背，靠背装席面软屉。靠背上方安竖向两枨，中装板，做云头开光，开光内镶嵌纹理优美的大理石。靠背上方搭脑另安，成大块面弧形，内凹外转，以配合头部曲线。后腿经由转轴至前方，向上延伸，略作折角，与修长扶手相接。折角处装曲边角牙起加固作用。扶手自靠背上方伸出，并转轴以方便开合，扶手曲线较为直挺，于扶手处略外展，出头较长。座面编织藤面，以配合折叠。

此种交椅式躺椅传世较为少见，在古画、笔记小说里有描绘。此件躺椅造型简洁，比例修长，气质文雅。

It's made of white dalbergia cochinchinensis of refined material with gorgeous veining and elegant gloss. The woven rattan backrest contrasts with wooden components as a clash of different materials, textures and colours.

As the central components of the structure, two pivots are installed in the middle of legs; front legs extend upward and are directly butted with armrests' rear ends to become vertical frames of the backrest; between the two vertical frames connects horizontal rails as the back edge of the seat and frames of the backrest; above the woven rattan back cushion installs two vertical rods to divide the upper frame into three sections respectively embedded with marble-inlaid plaques carved in cloud-shaped traceries; above the backrest installs a cambered headrest to perfectly fit the head curve; rear legs extend upward and are slightly curved then directly butted with long armrests above the seat; at the curved corners of rear legs installs corner apron plates as reinforcement; two slender and straight armrests with two front ends slightly bent outward, protrude a bit over the backrest and can pivot to fold; the woven rattan seat surface is foldable as well.

Such existing folding lounge chair is extremely rare and can be seen in ancient paintings or described in literary sketches. This chair is of clean and simple structure, slender silhouette and elegant beauty.

031

交椅式躺椅

Folding Lounge Chair

52×39×45/94cm

此躺椅白酸枝制，借用交椅结构，可折叠开合。

靠背四面攒框而成，靠背面心安藤面软屉，以承接靠背，增加舒适性。靠背之上另装异形搭脑，以承接头部。靠背两侧竖柱向下延伸与后腿交于转轴，落于地面。靠背偏上做转轴，伸出长直的扶手，于扶手处外展，扶手横截面为方形。

近扶手处装腿，腿向下延伸，与前腿交于转轴，落于地面。腿偏上部分安转轴，装座面。座面为四面攒边，面心装席面软屉，边抹做泥鳅背压边线。座面可绕转轴折叠。

四腿相交于转轴后与托泥连接，接榫处出长铜片加固，铜片两端做云头曲线装饰。

整件躺椅造型方正，构件横截面多为方材，给人方正规整的视觉感受。整件躺椅有三处转轴设计，可以通过转轴折叠。

原物载王世襄《明式家具研究》。

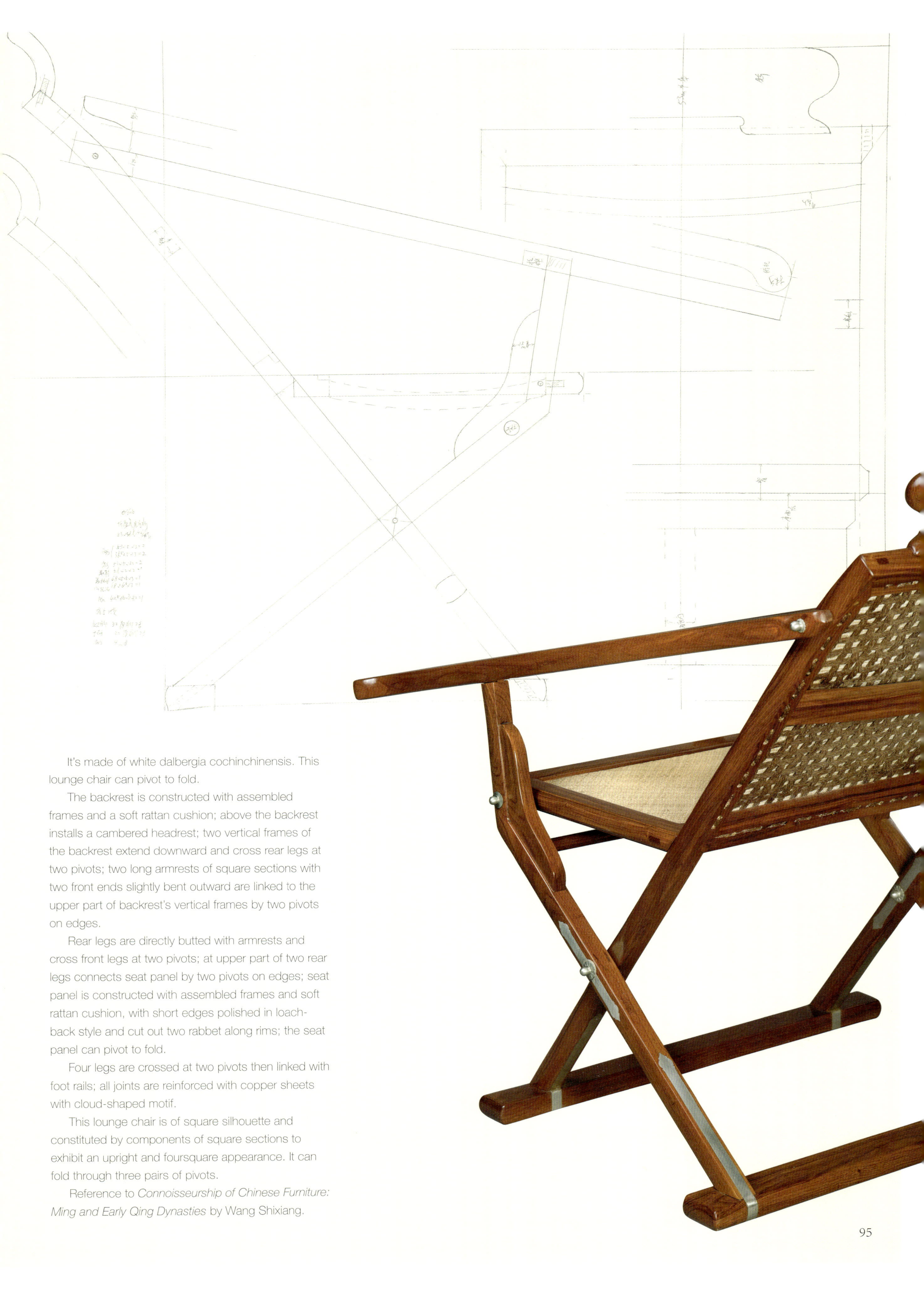

It's made of white dalbergia cochinchinensis. This lounge chair can pivot to fold.

The backrest is constructed with assembled frames and a soft rattan cushion; above the backrest installs a cambered headrest; two vertical frames of the backrest extend downward and cross rear legs at two pivots; two long armrests of square sections with two front ends slightly bent outward are linked to the upper part of backrest's vertical frames by two pivots on edges.

Rear legs are directly butted with armrests and cross front legs at two pivots; at upper part of two rear legs connects seat panel by two pivots on edges; seat panel is constructed with assembled frames and soft rattan cushion, with short edges polished in loach-back style and cut out two rabbet along rims; the seat panel can pivot to fold.

Four legs are crossed at two pivots then linked with foot rails; all joints are reinforced with copper sheets with cloud-shaped motif.

This lounge chair is of square silhouette and constituted by components of square sections to exhibit an upright and foursquare appearance. It can fold through three pairs of pivots.

Reference to *Connoisseurship of Chinese Furniture: Ming and Early Qing Dynasties* by Wang Shixiang.

032

禅椅

Zen Chair (for Meditation)

78×74×53/86cm

椅子由黄花梨制成，选料精良，色泽莹润。

椅子座面四面攒边，面心装席面软屉。边挺瘦薄，边抹做泥鳅背，不做线脚处理，更显得边挺细如边柱。座面之下装四腿，出挓，四腿穿过座面，向上延伸，分别与搭脑和扶手连接。四腿截面内方外圆，四腿上沿截面圆形，细节处理微妙。座面下，两腿之间安罗锅枨加矮老，形成错落空间，小空间与座面上的靠背和扶手大空间相呼应，错落有致，有韵律节奏。四腿下端安步步高赶枨，前面踏脚枨最矮，两侧略高，后面最高，又形成一组错落空间。

整件椅子以柱状构件勾勒轮廓为主，围合的空间空白处理，给人更多想象的空间，整件家具显得空灵剔透。

此种椅子在宋代古画中有见，多为文人雅士所坐，靠背不能依靠，只能正襟危坐，学者称其为“折背样”。

It's made of yellow *Huali (Huanghuali)* wood of refined material with smooth veining and elegant gloss.

The seat panel is constructed with assembled frames and soft rattan cushion, with thinner edges polished in loach-back style without any decorative lines or rabbets; below the seat installs four splayed legs of half-round-and-half-square sections, which extend upward through the seat and are respectively butted with headrest and armrests by having legs' upper ends processed into round sections; below the seat, hunched rails and short rods are butted between legs to divide the space and echo the silhouette of the backrest and armrests in rhythmic beauty; stretchers are linked between legs with varied height to also form a space of rhythmic change: the front foot rail is the lowest, and the back stretcher is the highest.

This chair is mainly outlined with cylindrical components, and leaves much space open for imaginations, therefore exhibiting an exquisite celestial beauty.

It can be seen in ancient Chinese paintings of Song Dynasty (960-1279), mostly used by literati who must keep their upper bodies straight as it has no backrest for them to rely on. It's also called "break-back style" by scholars.

033

镶玉西番莲纹太师椅（附方几）

Jade-inlaid Imperial Armchair Carved with Bas-relief of Passiflora Patterns (with a Square End Table)

75×56×48/106cm

太师椅由紫檀制成。靠背竖直，以横竖材攒框，横材使用双混面压边线的线脚处理，竖材使用西洋柱式。面心装板，浅浮雕西番莲纹饰。靠背搭脑高起，以双混面压线的线脚，作蜿蜒曲线，形成独特造型，正中镶玉，做圆寿纹。扶手再矮一截，亦攒框装板，板上浅浮雕纹饰。

座面做双层，边抹做双混面压边线处理。双层之间装绦环板，浮雕卷草纹饰。四腿做西洋柱式。腿下接托泥，托泥亦作双混面压边线处理。托泥下踩小足。

太师椅采用西洋做法，多用西洋柱式、西番莲纹饰装饰，又配以镶玉，整件座椅显得富丽堂皇，精致富贵。

几为配合太师椅而设计，几面亦作双层，边抹做双混面压边线处理。两侧之间装绦环板，上浮雕卷草纹。几面下接四腿，四腿做西洋柱式。四腿下接托泥，托泥边抹亦做双混面压边线处理。托泥之间攒成冰裂纹。托泥之下承小足。

It's made of red sandalwood. The vertical backrest is constructed with assembled frames: the horizontal components are of double cambered edges with rabbets along rims as borderlines; the vertical components are styled in western column; plaques are embedded and carved with bas-relief of passiflora patterns; backrest is raised in the middle as a headrest outlined with double cambered edges with rabbets along rims and embedded with a round jade plaque carved with abstract pattern of Chinese character "寿" (meaning longevity); vertical armrests stand lower and are also assembled and embedded with plaques of bas-relief.

The double-deck seat panel is constructed with double cambered edges with rabbets along rims; between the two decks is embedded with a decorative waist panel carved with arabesque bas-relief; four legs are styled in western column and butted with foot rails of double cambered edges with rabbets along rims; four small foot pads are installed below foot rails.

This imperial chair adopts several western designs including vertical components of western-column style and bas-relief of passiflora patterns, and is embedded with jade carving, which embellish this throne-like armchair with extreme magnificence and exquisite sumptuousness.

The matched end table (*Ji*) is designed in the same style, with double-deck table top constructed with double cambered edges with rabbets along rims; between the two decks is embedded with a decorative waist panel carved with arabesque bas-relief; below the table top installs four legs styled in western column and butted with foot rails of double cambered edges with rabbets along rims; between four foot rails installs multiple chips assembled in geometrical ice-crack pattern; four small foot pads are installed below foot rails.

承具

Furniture-to-present/carry

顾名思义，承具就是可以盛放器具的家具，主要有几、桌、案三种。几在席地而坐时期多指凭几或小型案，明清时期则多将长宽较小而高者名几，如香几、茶几、花几、架几均是此类。桌在古代本指短而阔者，今人根据造型，将四足位于四角者称为桌，也即常见的有束腰结构、四面平结构的承具，主要是从古代须弥座造型体系发展而来。而四足缩进者被归为案类，以夹头榫、插肩榫结构为代表，多狭而长，是从古代建筑梁架体系发展而来。古典家具中，承具是除了坐具以外第二丰富的门类，虽然其目的是提供一个工作、承接用的平台，但平面变化多端，腿足造型各异，高矮大小不同，形成了千变万化的样式。

As the name refers, furniture-to-present/carry indicates those objects which are used for bearing or presenting things, mainly including end table (*Ji*), table, and narrow table. During those periods when people sat on mats laid on the ground, the end table mostly referred to arm-prop or small stray table; In the Ming and Qing Dynasties, it referred to those furniture with smaller and higher table top, like incense table, tea table, flower table, narrow shelving table, etc.. In ancient time, table usually referred to the furniture with short and broad top panel; however, nowadays, those with four legs butted at four corners along table top edges are called table, including the common girdled waist structure and flat corner structure (corners butted with mitered corner bridle joints), which mainly originate from ancient Sumeru seat; and those with four legs installed within edges of the table top are called narrow table (*An*), including T-bridle joint structure (literally called clamp joint) and molded-frame bridle joint structure; narrow table is mainly in slender appearance and originates from the truss structure of ancient Chinese architecture. Furniture-to-present/carry is the second largest category in classical Chinese furniture after furniture-to-sit-in, with the purpose of providing a platform to work on or to hold. With various table tops, ever-changing legs' structures and different heights and sizes, it develops numerous styles and forms.

几

End Table (Ji)

几主要包括香几、茶几、花几、架几等，其中香几是古人供奉神祇、祈愿之用，既要体现参拜者虔诚的心情，又要体现对神祇的尊重，故而其造型往往最为卓越，也是传世家具中最有代表的一类。今人制香几，可以在室内随意搁置，以其亭亭玉立的造型点缀空间，若是搁置香炉焚香，烟雾袅袅，与香几相互映衬，形成一个雅意非凡的空间。其他的几则视需求而设。香几常见有圆形、方形、长方形、梅花形、海棠形、葵花形等造型，既有简练的四面平式造型，也有高束腰、三弯腿造型等。尤其是三弯腿造型的腿足，中部翻生花叶，末端卷转，有“蜻蜓腿”之称，是香几最为典型的造型特征。正是由于种种优点，在各类拍卖会中，香几往往是收藏家最为青睐的品种，其成交价格也往往是最高的一类。

The end table (*Ji*) mainly includes incense table, tea table, flower table, narrow shelving table, etc., among which the incense table used by the ancients during ritual ceremonies of offering sacrifices, worshiping and praying, is mainly built in styles or structures of the most solemn manner and celestial beauty to fully convey worshipers' conviction and their reverence for the god or deities. Nowadays, incense table can be placed in living space casually. Its slim and graceful appearance is an exquisite decoration for the room; or having an incense burner displayed on the table, with the rising curling smoke echoing the curvature of the end table, can easily create an exceptional space of tranquility and divinity. Other kinds of end tables shall be applied in accordance with different demands. Top panels of incense tables are commonly processed into round, square, rectangular, cinquefoil-shaped, begonia-shaped, chrysanthemum-shaped, etc.; they usually adopt structures of flat-corner (corners butted with mitered corner bridle joints), high girdled waist, and the most representative feature of three-curved legs (cabriole legs) (with middle part decorated with floral patterns and end part rolled in as "dragonfly's leg"), etc.; with these merits, in many auctions, incense tables are widely pursued by collectors and can usually achieve the highest hammer price.

034

卷球足带托泥长方香几

Rectangular Incense Table with Rolled-pearl Feet Butted with Foot Rails

46×40×77cm

方几由白酸枝制成，纹理细腻，色泽莹润。

香几正方，几面四面攒边，面心装硬板，四面边抹边缘起拦水线。边抹做混面压边线。几面下收束腰，束腰下接修长直挺的四腿，四腿向下略收细，内卷球收尾。四腿起挓明显，形成上小下大的稳定感。两腿之间安壶门牙条，壶门曲线翻转舒畅，与四腿连接，形成几面下空间。壶门牙条和四腿边缘不做阳线线脚处理，使得腿足更显光素修长。四腿之下承接托泥，托泥细薄，与四腿呼应，托泥下起弧线，切出四足，使托泥离开地面。

方香几各构件皆修长纤细，特别是四腿高挑秀美，下内卷球更是精巧有致，成就明式之经典。

It's made of white dalbergia cochinchinensis of refined material with smooth veining and elegant gloss.

The incense table has a rectangular top panel constructed with assembled frames outlined by embossed retaining lines, a hard central plaque, and cambered edges cut with a rabbet along the bottom rim; below the table top connects a girdled waist and four slim and straight legs slightly tapering downward and rolled up as pearls at end; four legs obviously lean outward to form a steady structure; between legs installs *Kun*-gate apron plates with smooth curves to divide a space below; without any borderlines or decorations makes its four legs appear to be more slender; below four rolled-pearl feet connects foot rails of slim and thinner style to echo four legs; the bottom side of foot rails are cut in with a curve to leave four foot pads at four corners to lift foot rails up from the ground.

This rectangular incense table is constituted by slim components, especially four slender legs, exhibiting a graceful and celestial beauty; four rolled-pearl feet are extremely exquisite; hence it is undoubtedly a piece of classical furniture of Ming style.

035

卷球足带托泥长方香几

Rectangular Incense Table with Rolled-pearl Feet Butted with Foot Rails

55×38.5×77cm

香几由黄花梨制成，黄花梨色泽温润，纹理丰富，油性十足。

香几几面正方，四面攒边，面心装硬板，四面边抹边缘略高起拦水线。边抹做混面压边线，混面竖直略圆，与整体简素的造型相呼应。几面下装束腰，束腰矮薄无饰。束腰下接四腿，腿修长直挺，出挓外撇明显，形成上小下大的稳定感，足部内卷成球。两腿之间装壶门牙条，壶门曲线兜转有力，自中心向两侧蜿蜒翻转，与修长的四腿曲线连接。四腿之下承接托泥，托泥用料纤细，下切出四足，使托泥离开地面。

此件方香几造型简洁素雅，四腿修长秀美，与壶门牙条曲线形成优雅的虚空间，造就此件香几空灵剔透的气质。

It's made of yellow *Huali (Huanghuali)* wood with elegant colour, rich and gorgeous veining and rich oiliness.

The incense table has a rectangular top panel constructed with assembled frames outlined by embossed retaining lines, a hard central plaque, and cambered edges cut with a rabbet along the bottom rim to echo the total construction of simple and clean beauty; below the table top connects a short unadorned girdled waist and four slim and straight legs slightly tapering downward and rolled up as pearls at end; four legs obviously lean outward to form a steady structure; between legs installs *Kun*-gate apron plates with full and dramatic curves running towards two sides to butt with the curvature of four legs; below four rolled-pearl feet connects foot rails of slim and thinner style; the bottom side of foot rails are cut in with a curve to leave four foot pads at four corners to lift foot rails up from the ground.

This rectangular incense table is of clean and simple structure; four slender legs and graceful curvature of *Kun*-gate apron plates outline an elegant void space below, and create an exquisite and ethereal beauty.

036

高束腰霸王枨方香几

Rectangular Incense Table with High Girdled Waist and Curved Diagonal Rails

50×50×78cm

香几以铁梨木制成，色泽沉稳，气质内敛。

几面攒框装板，四周边框做拦水线，可阻挡使用时溢出的汤水流下。边框侧面做成冰盘沿，造型简素，弧线饱满圆润。高束腰，一任简素。牙板微鼓出，沿边起线，牙板与腿足以柔婉的曲线相合，直腿足，末端内翻为卷足，为半个如意云头造型，与常见的马蹄腿和回纹腿皆不相同，富于新的变化，颇有意趣。腿足上设霸王枨，顶在几面下的穿带上，以保持结构的稳固。

此几属于有束腰结构中最为基本者，也是最为经典的制式，但高束腰的设计和特殊的卷足造型，带来新颖之感，颇耐观赏。

It's made of lignumvitae with dark and solemn colour, elegant and subtle gloss.

The top panel is constructed with assembled frames outlined by embossed retaining lines to stop the overflowing water, and unadorned cambered edges narrowed down widthways (also called ice-plate edge); below the top panel connects a high girdled waist; apron plates outlined with embossed borderlines slightly bulge and are butted with four legs in smooth curve; four legs are slim and straight, rolled up as half *Ruyi*-cloud feet at end instead of the common hoof feet or fretwork feet, with original changes and charm; between the upper part of four legs installs curved diagonal rails butted into lap joints at the bottom of the table top to maintain a steady structure.

This incense table with girdled waist is built in the most basic and classical structure, with the design of high girdled waist and special rolled-up half *Ruyi*-cloud feet to reveal its novelty and subtle beauty.

佚名《燕寝怡情》册　美国波士顿美术馆藏

Illustration of *Noblemen's Everyday Life* (album)
by anonymity
From Museum of Fine Arts, Boston

037

高束腰霸王枨带底座方香几

Rectangular Incense Table with High Girdled Waist, Curved Diagonal Rails and Bottom Base

50×50×86cm

此香几造型与前者接近，但是在其基础之上，增设有束腰台座，形成了迥然不同的效果，正可对比品鉴，一简一繁，各有优点。

以铁梨木制成，几面攒框装板，四周边框做拦水线。边框侧面做成冰盘沿，造型简素，弧线饱满圆润。高束腰，牙板微鼓出，沿边起线，牙板与腿足以柔婉的曲线相合，直腿足，末端内翻为卷足，为半个如意云头造型。腿足上设霸王枨，顶在几面下的穿带上，以保持结构的稳固。腿足下承有束腰内翻马蹄足台座。

This incense table is built in a similar structure with the previous except for the extra bottom base; juxtaposition of the two tables can clearly showcase their differences in style and functions.

It's made of lignumvitae. The top panel is constructed with assembled frames outlined by embossed retaining lines to stop the overflowing water, and unadorned cambered edges narrowed down widthways (also called ice-plate edge); below the top panel connects a high girdled waist; apron plates outlined with embossed borderlines slightly bulge and are butted with four legs in smooth curve; four legs are slim and straight, rolled up as half *Ruyi*-cloud feet at end; between the upper part of four legs installs curved diagonal rails butted into lap joints at the bottom of the table top to maintain a steady structure; a bottom base with girdled waist and four hoof feet is connected to four tips of legs.

038

霸王枨带托泥四面平长方香几

Rectangular Incense Table of Flat Corner Structure with Curved Diagonal Rails and Foot Rails

103×59.6×81cm

香几由红酸枝镶瘿木而成。四面平，几面和腿之间使用综角榫连接，面心装纹理优美的瘿木木心，花纹细密旋转，玄妙多姿，绮丽可爱。橙黄的瘿木色彩和深色的家具构件，形成鲜明的对比，相映成辉。

几面之下接四腿，四腿截面方形，末端内翻马蹄收尾。四腿偏上位置伸出霸王枨，以勾挂榫与腿连接，霸王枨曲线劲挺，上扬，收于桌面底部的穿带上。

四腿之下安托泥，托泥四面攒边，边抹做混面压边线线脚，四角安小足，足部做曲线修饰。

此件四面平长方香几，采用独特的四面平结构，霸王枨和内翻马蹄的修饰，使得整件家具曲线优雅、简洁而又细腻，气质不凡。

It's made of red dalbergia cochinchinensis; table top is framed with flat corner structure (corners butted with mitered corner bridle joints, also called cubic-cone rice-pudding joint) and embedded with a burl panel of swirls of gorgeous veining; the orange burl panel and other dark components create a striking contrast.

Below the table top connects four legs with square sections and hoof feet at end; between the upper part of four legs installs curved diagonal rails with crook joints which rise up and are butted into lap joints at the bottom of the table top.

Foot rails are installed below foot tips, with assembled frames and cambered edges cut in a rabbet along the bottom rims; four small foot pads carved with decorative curves are butted into bottom corners.

This rectangular incense table is built in a unique flat corner structure and adorned with curved diagonal rails and hoof feet to outline a clean and simple silhouette of exquisite elegance.

039

霸王枨卷球足带托泥四面平长方香几

Rectangular Incense Table of Flat Corner Structure with Curved Diagonal Rails and Rolled-pearl Feet Butted with Foot Rails

82×48×81cm

香几由白酸枝制成。四面平，几面四角由两边抹和腿以综角榫连接，面心装硬板，平装。几面厚度与腿足宽度相当，形成一致的修长的体量感。几面下接四腿，腿修长直挺，末端内卷球收尾。

四腿偏上位置装霸王枨，与腿勾挂榫连接。霸王枨成S形曲线，翻转向上，与几面下穿带连接。

腿下接托泥，托泥四面攒边，竖直线脚，不做任何修饰，与四面平的几面处理方式相同。托泥下切出小足，将托泥与地面分离，可以防潮隔湿。

此香几采用独特的四面平做法，四腿直接与几面相接，并用霸王枨加固，加之四腿之下的托泥固定四腿，使得各构件稳定坚固，形成流畅的方形空间，如雕塑般秀美俊雅。

It's made of white dalbergia cochinchinensis; table top is framed with flat corner structure (corners butted with mitered corner bridle joints, also called cubic-cone rice-pudding joint) and embedded with a flat hard panel, with an edge width same as leg width to create a consistent appearance of slenderness; below the table top installs four slim and straight legs rolled up as pearls at end.

Between the upper part of four legs installs s-shaped curved diagonal rails with crook joints which rise up and are butted into lap joints at the bottom of the table top.

Foot rails are installed below foot tips, with assembled frames and unadorned straight edges to echo the flat table top; four small foot pads are carved out at bottom of foot rails at corners to separate the table from the ground due to humidity.

This rectangular incense table is built in a unique flat corner structure; four legs are directly butted into the table top and reinforced with curved diagonal rails; foot rails are attached below four feet; the whole structure guarantees its steadiness and also outlines a flowing square space of sculptural beauty.

040

海棠开光带托泥委角长方香几

Rectangular Incense Table with Dented Corners and Foot Rails Decorated with Hollowed Begonia-shaped Tracery

76.5×38×80cm

长方香几由紫檀镶瘿木制成，紫檀沉穆，瘿木淡雅，相得益彰。

几长方，几面四面攒边，面心装瘿木，花纹细密缠绕，如满架葡萄，几面四角做委角处理。浅色瘿木和深色紫檀形成鲜明对比，又协调一致。几面下接束腰，束腰略高，四角亦做委角处理，束腰之上开海棠形长开光，开光下边缘正中向上翻出云头，直顶上边缘。开光边缘起一圈阳线，以勾勒轮廓。

束腰之下接四腿，腿修长直挺，末端以内翻马蹄收尾。腿背之上亦做委角处理，与上面束腰和几面一致。

整件家具最大的特色是从几面、束腰至腿、托泥和小足一贯而下的委角处理，上下贯通，一气呵成。束腰之上的海棠卷云头开光，使得香几更加通透空灵。

It's made of red sandalwood with dark and solemn colour and embedded with a burl panel of elegant veining.

The incense table has a rectangular top panel constructed with assembled frames with dented corners and embedded with a burl plaque of swirls of refined veining as vines of grapes; the light-coloured burl and dark red sandalwood create a striking yet harmonious contrast; below the table top connects a high girdled waist with dented corners and hollowed begonia-shaped tracery outlined with embossed borderlines; a delicate carving of cloud is clamped right in the middle of each long begonia-shaped tracery.

Below the girdled waist connects four slim and straight legs rolled up as hoof feet at end; four legs are also processed to have dented shoulders to consist with the same design on the above girdled waist and table top.

This incense table is quite distinct with its consistent dented-corner design on table top, girdled waist, leg shoulders, foot rails and foot pads; the hollowed begonia-shaped tracery on table waist gives the furniture a celestial beauty.

041

海棠开光带托泥委角长方香几

Rectangular Incense Table with Dented Corners and Foot Rails Decorated with Hollowed Begonia-shaped Tracery

81×38×80cm

香几由黄花梨制成。长方形，几面四面攒边，面心平装硬板。几面四角委角，丰富线脚变化。边抹做混面压边线线脚处理。几面下接束腰，束腰四角也做委角处理，束腰上开两海棠形开光，开光下边缘正中间向上翻出云头，海棠开光起阳线以勾勒边缘。

束腰下接四腿，腿直挺向下，末端内翻马蹄收尾。腿背做委角处理，与上部的束腰和几面委角相一致，委角从上到下一直延伸到马蹄足，以及托泥和小足。两腿之间装直牙条，牙条和腿外缘做阳线处理。

四腿之下安托泥，托泥四面攒边，边抹做混面压边线处理，四角委角。托泥下接小足，足部也做委角处理。

此件长方香几最独特之处在于从桌面四角向下延伸至托泥小足的委角处理，以及束腰上的海棠卷云头开光，让这件长方香几线脚丰富，造型别致，秀雅可爱。

Made of yellow Huali (*Huanghuali*) wood, the incense table has a rectangular hardwood top panel constructed with assembled frames with dented corners and various decorative mouldings, and cambered edges cut in a rabbet along the bottom rims; a girdled waist is attached below with dented corners to coincide with the structural style of top panel, and carved out a hollowed begonia-shaped tracery outlined by embossed borderline and ardorned by a cloud-shaped centerpiece on each side.

Below the girdled waist connects four slim and straight legs rolled up as hoof feet at end; four legs are also processed to have dented shoulders to coincide with the same design on the above girdled waist and table top, which run down vertically to appear as grooves on the exterior of legs, hoof feet, foot rails and foot pads; unadorned straight apron plates are butted between legs and borderlined by embossed lines to emphasize its contour.

Foot rails are installed below foot tips, with assembled frames, dented corners and cambered edges cut in a rabbet along the bottom rims; four small foot pads with dented corners are butted into bottom corners.

This incense table is quite distinct with its consistent dented-corner design on table top, girdled waist, leg shoulders, foot rails and foot pads; the hollowed begonia-shaped tracery on table waist gives the furniture a celestial beauty.

042

卷草纹三弯腿带托泥方香几

Square Incense Table with Arabesque Bas-relief, Three-curved Legs (Cabriole Legs) and Foot Rails

42.5×42.5×98cm

方香几由白酸枝制成，木材纹理优雅细腻，色泽莹润，油性充足。

香几正方，几面四面攒边，面心由木条攒成几何纹样。边抹边缘略抬起阳线成拦水线。边抹起冰盘沿压边线线脚。几面下装略高束腰，束腰上下起略高似托腮的台，束腰中间起双鱼门洞开光，开光边缘起阳线，以勾勒轮廓，开光细长，增加空间通透感。

束腰下安三弯腿，腿S形曲线，自束腰上鼓起，顺势而下收起，偏上端起卷叶曲线，于足端外翻出卷球，兜转有力，一气呵成。两腿之间装壶门牙条，壶门曲线起伏较大。壶门边缘起阳线，阳线顺势向上翻转，卷成浅浮雕的卷草纹，卷草翻转自如。壶门牙条与腿足连接处亦雕刻卷草纹。壶门牙条的边缘起阳线，与腿足起的阳线连接，腿部上端翻转卷叶的部分，阳线向腿内翻折出与卷草相呼应的阳线纹饰。

腿足之下承托泥，托泥四面攒边，边抹做混面压边线，托泥四角下安小足。

方几大弧度的三弯腿兜转自然有力，成为此件家具最大的造型特色。

It's made of white dalbergia cochinchinensis with elegant and smooth veining, subtle gloss and rich oiliness.

The square top panel is constructed with assembled frames outlined by embossed retaining lines to stop the overflowing water, a central plaque of geometrical patterns pieced by chips and splinters, and unadorned cambered edges narrowed down widthways (also called ice-plate edge) and cut in a rabbet along the bottom rims; below the top panel connects a high girdled waist with two stepped mouldings on up and bottom sides, and carved with hollowed slim double-fish traceries outlined with embossed borderlines in the middle areas to enhance a sense of transparent.

Below the girdled waist installs s-shaped three-curved legs (cabriole legs) bulging out below the waist and then running inward lengthways with smooth floral curves at the upper part and rolled-up pearl feet at end; between legs connects *Kun*-gate apron plates outlined with embossed borderlines with dramatic curves which run upward in the middle and turn into graceful arabesque bas-relief; at the joints of apron plates and legs is carved with intaglios of floral patterns as decoration; four legs are butted with apron plates along embossed borderlines and adorned with floral curves at the upper part to echo the arabesque bas-relief on apron plates.

Foot rails are installed below foot tips, with assembled frames and cambered edges cut in a rabbet along the bottom rims; four small foot pads are butted into bottom corners.

Its dramatic three-curved legs (cabriole legs) absolutely distinguish this incense table for a unique beauty of strength and elegance.

043

卷草纹三弯腿带托泥三足圆香几

Round Incense Table with Arabesque Bas-relief,
Three Three-curved Legs (Cabriole Legs) and Foot Rails

38×38×95.5cm

此香几由白酸枝制成，白酸枝色泽雅致，纹理流畅自然。

圆几攒成圆框，面心打槽装板。边抹做冰盘沿压边线线脚处理。几面下接束腰，束腰高于几面厚度。

束腰下接三腿，腿三弯，自束腰之下鼓出，在向内收敛后，又向外翻出，上雕卷叶。束腰下、三腿两两之间安壸门牙条，牙条与三腿交圈膨出，壸门曲线兜转有力，自中心向上翻转，做浮雕卷草纹，两侧与腿连接。牙条之上起阳线，一直延伸至腿上部，继续向内翻折出卷叶纹饰。牙条与腿相接处再浅浮雕云头纹饰，似镶钉在转角处的铜包角。

三腿之下安圆形托泥，托泥边抹做素混面处理，混面饱满圆润，与几面遥相呼应。托泥之下装三足。

此香几体形颀长，S形腿亭亭玉立，秀雅别致。

明万历
《元曲选图》插图

Wanli Period of Ming Dynasty (1573-1620)
Illustration of *Selected Illustrations of Verses of the Yuan Dynasty*

It's made of white dalbergia cochinchinensis with elegant colour, natural and smooth veining.

The round top panel is constructed with assembled frames, an embedded round central plaque, and unadorned cambered edges narrowed down widthways (also called ice-plate edge) and cut in a rabbet along the bottom rims; below the top panel connects a high girdled waist with its height wider than the width of the table top edge.

Below the girdled waist installs three three-curved legs (cabriole legs) bulging out below the waist and then running inward lengthways with smooth floral curves at the upper part and bent-out feet carved with a rolled-leaf at end; between legs connects bulging *Kun*-gate apron plates with dramatic curves which run upward in the middle and turn into graceful arabesque bas-relief; three legs are butted with apron plates along embossed borderlines and adorned with floral curves at the upper part; at joints of apron plates and legs are carved with cloud bas-relief as well, resembling copper corner protectors.

Foot rails are installed below three foot tips, with unadorned cambered edge to echo the clean and simple style of table top; three small foot pads are butted into rail bottom.

This incense table is built in slender structure with s-shaped legs, exhibiting an exquisite beauty of elegance.

044

三弯腿带托泥五足圆香几

Round Incense Table with Five Three-curved Legs (Cabriole Legs) and Foot Rails (a Pair)

41×41×97cm

香几成对，由白酸枝制成，纹理流畅，色泽温润。

香几几面由五段弧形木边攒成，面心平装硬板。木边边缘略高起做拦水线，边抹做打洼沿压边线线脚处理。几面下安束腰，束腰略高，上起海棠形阳线开光。束腰之下安托腮，托腮做打洼沿压边线线脚处理，与几面边抹处理一致。

托腮之下接五腿，五腿三弯挺劲，自托腮下开始膨出，顺势直下，自末端外翻出卷叶抱球，下再踩方形上小下大小座。五腿两两之间安壶门牙条，壶门曲线兜转有力，与腿连接。壶门曲线边缘起阳线，与腿连接后，阳线向腿背上翻转，似倒挂壶门。

五腿之下安托泥，托泥亦由五段弧形木边攒成，面心平装硬板。边抹做冰盘沿压边线线脚处理。托泥之下安小足。

此件香几最独特之处在于起腿足的处理，异于一般弯度较大的S形三弯腿，而是内敛劲挺的腿部曲线。整件香几造型饱满稳定，内敛静雅，文人气质浓郁。

This pair of round incense tables are made of white dalbergia cochinchinensis with smooth veining and elegant colour.

The round top panel is constructed with a frame assembled by five sections and outlined by embossed retaining lines to stop the overflowing water, an embedded flat hard central plaque, and an edge planed down into low-lying surface with a rabbet along the bottom rim; below the top panel connects a high girdled waist with bas-relief of begonia-shaped traceries; a stepped moulding below the waist is also planed down in low-lying surface to consist with the edge of table top.

Below the stepped moulding installs five three-curved legs (cabriole legs) with dramatic curves bulging out from the top, running slightly inward lengthways and tilting out as rolled-pearl feet at end; five feet touch on five small square foot pads; between legs installs *Kun*-gate apron plates outlined with embossed borderlines with dramatic curves which run across the exterior of legs and turn into upside down *Kun*-gate bas-relief.

Below five square foot pads installs a round foot rail assembled by a five-section frame and a hard central plaque, with cambered edge narrowed down widthways (also called ice-plate edge) and cut in a rabbet along the bottom rims; five small foot pads are butted into bottom of the round foot rail.

This incense table is quite distinct with design of its legs; varied from other common s-shaped three-curved legs (cabriole legs), five legs of this table are processed into subtle and masculine curves; The whole structure is steady and fully unfolded, exhibiting a tranquil and elegant aesthetics of ancient Chinese literati.

045

卷叶纹三弯腿带底座五足香几

Round Incense Table with Rolled-leaf Bas-relief, Five Three-curved Legs (Cabriole Legs) and a Bottom Base (a Pair)

48×48×118.5cm

香几成对，由白酸枝制成，色泽雅致，纹理优美。

香几几面由五段弧形木边攒成，各段木边相接处做委角，成梅花形，边抹做混面压双边线线脚处理。几面下安托腮和束腰，托腮置于束腰上下，做混面压边线线脚处理。束腰随着几面成梅花式，于委角处安矮老，矮老之间装绦环板，绦环板上开海棠形开光，开光边缘起一圈阳线，以勾勒轮廓。

束腰和托腮之下安五腿，五腿S形三弯，自托腮下膨出，顺势而下，向内收敛，偏下收敛之处翻折出草叶，末端向外翻转成云头，上卷草叶，下踩圆球。五腿外背依上面委角之势，亦做委角处理。托腮之下、五腿两两之间安壶门牙条，壶门曲线翻转有力，与腿连接。牙条和腿边缘皆起阳线，以勾勒轮廓。

五腿之下不安托泥而设台座式底座，台座与几面处理方式相似，亦为梅花式台面、托腮和束腰，束腰之上开壶门长开光，两端又辅以云头开光，边缘皆起阳线。托腮之下装膨起牙条，牙条上做云头曲线处理，并起阳线。如此叠制而做的台座，与几面呼应，形成造型完整一致的整体。

此对香几独特之处在于蜿蜒的S形腿足处理，以及比托泥稳重的台座设计。整件香几稳重一体，气宇非凡。

This pair of round incense tables are made of white dalbergia cochinchinensis with gorgeous veining and elegant colour.

The cinquefoil top panel is constructed with a frame assembled by five sections and dented at each joints to form a plum-blossom (five petals) shape; panel edge is cambered and cut in two rabbets along the top and bottom rims; below the table top installs a girdled waist clamped by two stepped moudlings with cambered edges and rabbets along the rims; waist is also dented at five joints in accordance with the table top to form a cinquefoil section, and is installed with five short rods at five dented joints; five curved decorative plaques carved with begonia-shaped traceries are butted in between short rods; and all five slim traceries are outlined by embossed borderlines.

Below the girdled waist and stepped moulding installs five s-shaped three-curved legs (cabriole legs) bulging out from the top, running slightly inward lengthways and tilting out as rolled-leaf feet with cloud bas-relief at end; five feet touch on five small pearl foot pads; on exterior of legs runs a dented rabbet lengthways in accordance with dented corners of the table top; below the stepped moulding installs *Kun*-gate apron plates outlined with embossed borderlines with dramatic curves between legs.

A bottom base is attached below five rolled-leaf feet instead of foot rails; the base is processed into the same style as the table top: it's constructed with a cinquefoil base top, a five-sectioned girdled waist carved with a slim *Kun*-gate tracery accompanied by two hollowed cloud patterns on two sides in each section and a stepped moulding; all the decorative carvings are outlined with embossed borderlines; below the moudling connects bulging apron plates carved in cloud-shaped curves and outlined with embossed borderlines; This bottom base with multi-layered structures echoes with the table top to maintain a unified style.

This pair of incense tables are quite distinct with their s-shaped legs and the bottom base with more complicated structures than the common foot rail, showcasing a dignified solemnity.

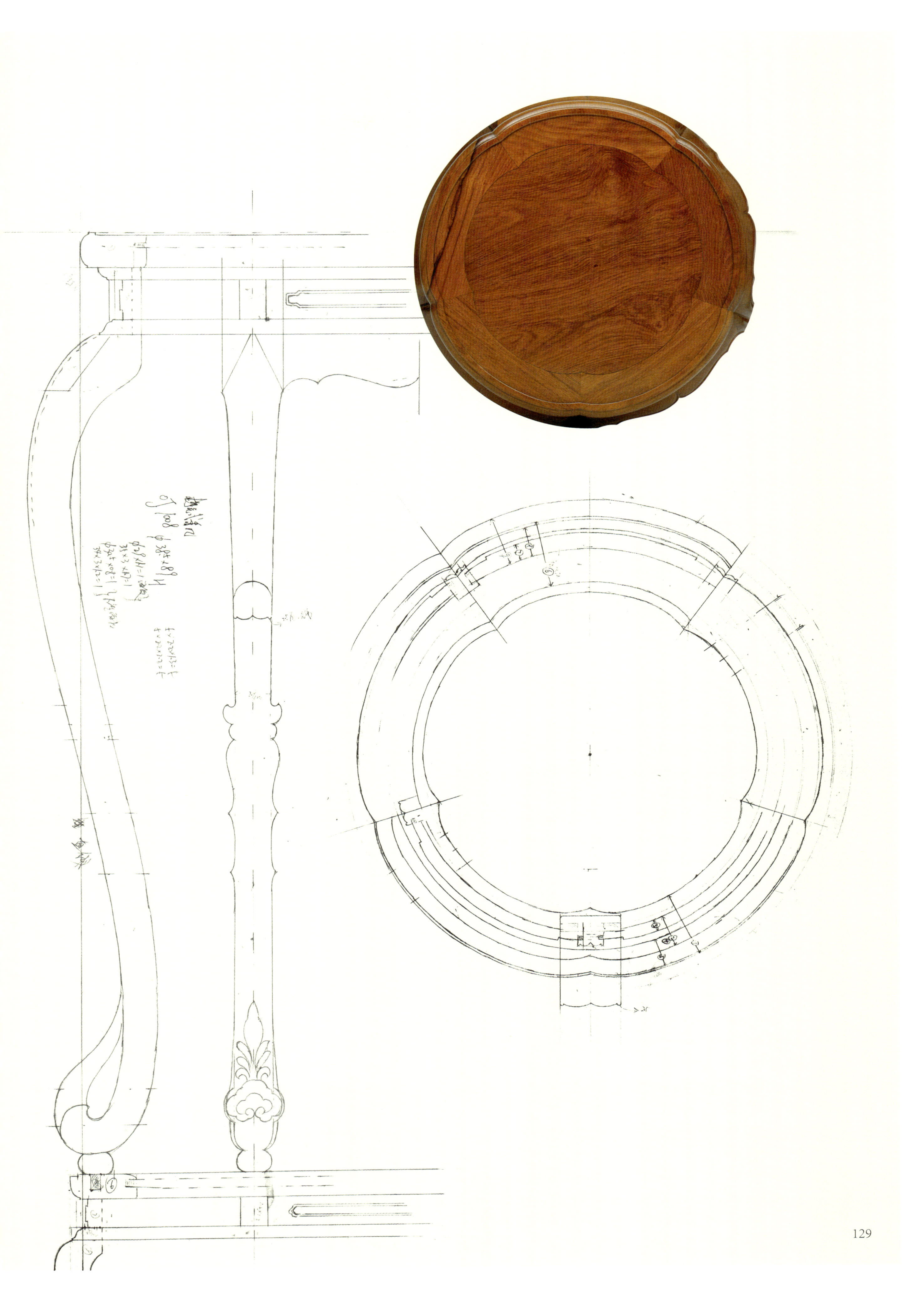

046

西番莲纹高束腰展腿式带底座长方香几

Rectangular Incense Table with High Girdled Waist,
Bent Legs, Bottom Base, and Bas-relief of Passiflora Patterns

50×40×97cm

香几四方，紫檀制成。几面四面攒边，面心平装板。边抹边缘高起做拦水线，边抹立面做浅浮雕委角开光，开光外雕卍字锦地，开光内雕西番莲纹饰。几面之下装高束腰和仰覆莲托腮，束腰之上开委角长开光，开光内透雕西番莲拐子纹。托腮之下做披肩式牙子，下装三弯腿，腿末端向外翻出云头。

四腿之下装台座式底座，与几面遥相呼应。台面四面攒边，面心板浅浮雕西番莲纹饰，外圈做委角边框。边抹做冰盘沿压边线线脚处理。台面下收束腰。束腰下装披肩式牙子，牙子上浅浮雕卷草纹饰，边缘起阳线勾勒轮廓。

此香几满雕西番莲纹饰，腿足三弯样式为典型的清式腿。腿足下设台座，异于一般的托泥，更显稳重大气。

The rectangular incense table is made of red sandalwood. The table top is constructed with assembled frames outlined by embossed retaining lines and a hard flat central plaque, with four straight edges carved with bas-relief of passiflora patterns framed by a rectangular window with dented corners against a background of Buddhist swastika motif; below the table top connects a high girdled waist clamped by two lotus mouldings on top and bottom sides; the waist is carved with openwork of passiflora and abstract dragon patterns framed by a rectangular window with dented corners on each side; below the lotus moulding installs cape-styled apron plates and three-curved legs (cabriole legs) with bottom ends rolled outward as cloud-shaped feet.

A bottom base is attached below four feet to echo with the table top: it's constructed with assembled frames and a central plaque carved with bas-relief of passiflora pattern and framed by a rectangular window with dented corners, with edges narrowed down widthways (also called ice-plate edge) and cut in a rabbet along the rim; below the bottom base connects a girdled waist and cape-styled apron plates carved with bas-relief of floral patterns and outlined with embossed borderlines.

This incense table is fully covered by carvings of various passiflora patterns; its three-curved legs (cabriole legs) is representative in the Qing Dynasty (1644-1911); the bottom base attached below four feet is quite different from the common foot rails, showcasing a dignified solemnity.

047

带托泥委角六边形香几

Hexagon Incense Table with Dented Corners and Foot Rails

45×45×100cm

香几六面委角花形，紫檀制成。几面由六段木边攒成，面心平装板。边抹边缘向上略高起做拦水线，边抹做打洼沿压双边线线脚处理。

几面下安高束腰，束腰也随几面做委角处理。束腰之下装托腮，托腮之下装六腿，腿修长竖直，不做曲线处理，末端翻马蹄。腿边缘起阳线，中做委角，并起双炷香阳线。两腿之间安透雕拐子纹牙条。

六腿之下安托泥，托泥随形，亦做六面委角花形。边抹做混面压边线线脚处理。托泥之下安小足，小足亦做委角处理。

此件香几体形修长，做六面委角花形，此造型自几面向下一直延续至托泥和小足，形成一贯的设计元素，使得整件家具整体一致。为典型的清式香几造型。

This incense table made of red sandalwood, has a flower-like (with dented corners) hexafoil table top constructed by six-sectioned assembled frames outlined by embossed retaining lines and a flat hard central plaque, with edges planed down into low-lying surface with two rabbets along the top and bottom rims.

Below the table top installs a high girdled waist with six dented corners to consist with the top panel; below the waist connects a moulding and six slim and straight legs rolled inward as hoof feet at end; legs are outlined with embossed borderlines and carved in a dented groove lengthways in the middle to echo the hexafoil-shape of table top, with double embossed lines running along the groove (also called incense line); between legs installs apron plates carved with openwork of abstract dragon patterns.

Foot rails butted together as a same flower-like (with dented corners) hexagon frame are attached below six feet, with cambered edge and a rabbet along the rim on each rail; six small foot pads with a dented groove carved in the middle are installed under foot rails.

This incense table is of slender appearance and characterized by its delicate flower-like hexafoil design applied from table top all the way down to foot rails and foot pads to maintain a unified style, as a representative classical incense table of Qing Dynasty (1644-1911).

048

高束腰带底座六边形香几

Hexagon Incense Table with High Girdled Waist and a Bottom Base

71×47.5×91.5cm

香几为六边六腿式，由白酸枝制成。几面为六段木边攒成，面心平装板，边抹做冰盘沿压边线线脚处理。几面下安高束腰，束腰上六角的位置安如矮老的短柱，短柱之间嵌装绦环板。绦环板上开海棠式开光，开光边缘起阳线，内做似攒斗式卍字锦，实为独板绦环板透雕而成。

束腰之下安托腮，托腮之下装六腿，腿自托腮顺势而下，略向内收，自末端外翻成卷球和草叶，下再踩球。托腮下两腿间安壸门牙条，牙条用料宽，壸门曲线兜转有力，两侧与腿相接。牙条和腿边缘皆起阳线。

六腿之下设台座式底座，较几面处理略简易，以分别主次。台面六边攒边，边抹做冰盘沿压边线。台面下安矮束腰，束腰下装六小腿和壸门牙条。

此六面香几体量较大，下设台座与几面相呼应，更显香几稳定持重。

This hexagon incense table made of white dalbergia cochinchinensis has a hexagon table top constructed by six-sectioned assembled frames and a flat hard central plaque, with edges narrowed down widthways (also called ice-plate edge) and cut in a rabbet along the bottom rim. Below the table top installs a high girdled waist with six short rods butted in its six corners, and six decorative plaques embedded between short rods; each plaque is carved (instead of being assembled) with a Buddhist swastika tracery framed by a slim begonia-shaped window with four dented corners out of an one-piece material.

Below the waist connects a moulding and six slim legs running slightly inward and tilting out as rolled-pearl-and-leaf feet at end; six feet touch on six small pearl foot pads; between legs installs broader *Kun*-gate apron plates outlined with embossed borderlines with dramatic curves.

A bottom base is attached below six feet, with unadorned simple structure to contrast with the heavily decorated table top; the base top is constructed with six-sectioned assembled frames and cambered edges narrowed down widthways (also called ice-plate edge) with a rabbet along the bottom rims; below the base top installs a short girdled waist and *Kun*-gate apron plates between six short legs.

This hexagon incense table is built in bigger size, with a bottom base to echo the table top, exhibiting a dignified solemnity.

桌

Table

桌本写作"卓""棹"等，顾名思义，有卓尔而立之意。桌是高型家具体系所有，目前所知大约出现在唐中期，敦煌莫高窟第85窟晚唐《楞伽经变》中屠夫杀生场面中的两张方桌，是较早的桌的形象，其造型为桌面下设四直足，其他并无连接或支撑作用的枨子、牙板等，显示出桌在初始状态下结构不够科学合理。在古人的概念里，卓然而起的高型承具，凡短小者，均名为桌，今人则根据结构，将腿足位于四角者归为桌类。依照尺寸和造型，桌可分为方桌、长方桌（即条桌）、圆桌、半圆桌、梅花桌等。其中方桌又根据尺寸大小分为八仙桌、六仙桌等，四面坐人，使用便捷，南方又称为"方台"。还有供学习或书画活动的桌，则名画桌，以刀牙板样式和四面平样式最为经典，尺寸可大可小，搭配官帽椅或灯挂椅，是上好的书房家具组合。

Table (" 桌 "), originally written as " 卓 " or " 棹 ", as the name refers, means standing high and straight, or outstanding as an extended meaning. Table belongs to the high-seat system, is known to appear in the mid-Tang Dynasty (618-907). According to the mural painting of *Story of Lankāvatāra Sūtra* in Mogao Grottoes (cave 85, late Tang Dynasty, 618-907), two square tables are depicted in a scene where a butcher is slaughtering animals, which are table's early appearance in ancient China, with four straight legs and a table top, without any jointing or supporting elements like rails or apron plates, therefore it reveals the table of early time might be built in a less scientific structure. In the concept of the ancient, among all the furniture-to-present/carry in the high-seat system, one can be called as table as long as it is small or short in size. Nowadays, according to the structure, those with four legs butted at four corners along table top edges are called table. In terms of size and form, tables can be categorized into square table, rectangular table (also called narrow table), round table, half-round table, cinquefoil table, etc.; among which square table can be further divided into large-sized square table for eight people (or table of eight-immortal) and medium-sized square table (or table of six-immortal), etc. in term of size. The square table can serve people on four sides, and is also called "square platform" by people in the south China. Moreover, the table used for study, calligraphy and painting activities is called drawing table which is typically constructed with T-bridle joint (clamp joint) and flat-corner structure (corners butted with mitered corner bridle joints) in various sizes; it can be displayed with yoke (official's hat)-back armchairs or yoke (lamp hanger)-back chairs, as a high-class furniture setting for study.

049

罗锅枨小方桌

Small Square Table with Hunched Rails

56×56×77cm

桌正方，紫檀镶瘿木而成。桌面四面攒边，面心镶嵌瘿木，瘿木色泽橙黄，花纹瘿节满密，灵动活泼。边抹做混面压双边线，混面饱满圆润。

桌面之下安四圆腿，四腿八挓，形成下大上小的稳定感。桌面下，两腿间，先安刀板牙子，牙子边缘起阳线，以勾勒轮廓。紧挨着刀板牙子之下，安高拱罗锅枨，罗锅枨高拱部分直接紧贴在刀板牙子之上。刀板牙子和罗锅枨双重加固的结构，让整件家具更加坚固稳定，刀板牙子和罗锅枨的搭配使用，又成为独特的设计元素，成为此件家具的点睛之处。

此件方桌为典型的梁柱式结构，造型简洁精炼，结构严谨坚固，实现了造型与结构的完美结合。

The square table made of red sandalwood has a table top constructed with assembled frames and an embedded burl central plaque with vibrant bright orange colour and swirls of refined and smooth veining; the full and round cambered edges are cut in two rabbets along top and bottom rims.

Below the table top installs four splayed legs with round sections to form a steady structure. Apron plates outlined with embossed borderlines are butted between legs. Below apron plates installs hunched rails; and hunched areas are directly butted with the above apron plates. Such double reinforced structure (with hunched rails and apron plates butted together) can greatly enhance the table's sturdiness and durability, and also appears as a striking design of originality.

This square table is built in typical girder-pillar structure, with clean and simple form, precise and sturdy structure, as a perfect balance between form and structure.

050

螭龙拐子纹马蹄腿小方桌

Small Square Table with Openwork of Abstract Dragon Pattern and Hoof Feet

69.5×69.5×81cm

桌正方，黑料红酸枝制成。桌面四面攒边，面心平装硬板，边抹做冰盘沿压边线线脚处理。桌面下收束腰，束腰上开并列的长鱼门洞开光，开光边缘起阳线勾勒轮廓。

束腰之下接四腿，腿竖直刚劲，末端内翻马蹄收尾。两腿之间安直牙条，牙条和腿的边缘接起阳线。牙条之下安花牙子，上透雕螭龙和拐子相互缠绕翻转，灵动活泼。

此桌为典型的束腰式结构，束腰上开的长条鱼门洞开光，以及牙条下的透雕花牙子，营造出空灵通透的空间感，减轻上部桌面的厚重感，增加稳定性。

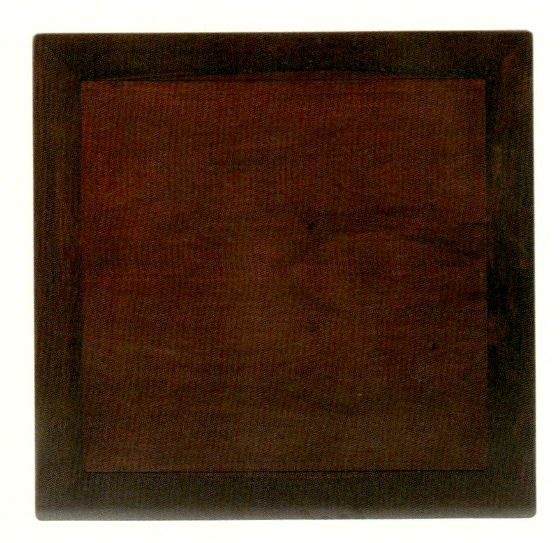

The square table made of red dalbergia cochinchinensis with black colour, has a table top constructed with assembled frames and a hard central plaque, with edges narrowed down widthways (also called ice-plate edge) and cut in a rabbet along the bottom rim as the borderline; below the table top connects a girdled waist carved with a pair of hollowed tiny slim long-fish-doorway windows outlined by embossed borderlines on each side.

Four straight and sturdy legs are butted below the waist and turned into introverted hoof feet at end; between legs installs straight apron plates outlined with embossed borderlines and butted with decorative panels carved with exquisite openwork of dragons intertwining with geometrical fretwork motif.

It's built in a typical girdled waist structure; hollowed slim long-fish-doorway windows and decorative panels carved with openwork of dragon patterns create an ethereal and transparent sense of space to significantly reduce the heaviness of the upper part and reinforce the steadiness.

051

灵芝卡子花罗锅枨方桌

Square Table with Ganoderma (*Ruyi*)-shaped Clamp Chips and Hunched Rails

89×89×82cm

方桌由黄花梨制成，选料上乘，纹理流畅自然，色泽莹润。

方桌桌面四面攒边，面心平装，边抹做冰盘沿压边线处理，边线较一般略宽，且做打洼处理。

桌面下安四圆腿，与桌面高低榫连接。桌面之下，两腿之间安罗锅枨，罗锅枨截面做素混面处理，混面圆润饱满，罗锅枨曲线流畅有力。桌面下、罗锅枨之上安两朵如意云头形灵芝纹卡子花，卡子花做花形，内反转出云头纹饰，雕琢精致，成为此件方桌的点睛之处。

罗锅枨和卡子花经常搭配使用，安在桌面和两腿之间，以加固并分界空间。此件方桌的罗锅枨和卡子花的搭配，恰是整件家具的独特之处。

It's made of yellow *Huali (Huanghuali)* wood of selected refined material with smooth, natural veining and elegant gloss.

Its square table top is constructed with assembled frames and a flat central plaque, with edges narrowed down widthways (also called ice-plate edge) and cut in a rabbet along the bottom rim as the borderline; typically, the rabbet is wider than the usual and is gouged out a rounding low-lying groove.

Below the table top installs four round-section legs by uneven double tenons; unadorned hunched rails with polished cambered surfaces and smooth and masculine curves are butted into upper parts of four legs; two ganoderma (*Ruyi*)-shaped clamp chips are installed between hunched rail and table panel on each side; clamp chips are exquisitely carved in abstract design of flower and cloud, as a striking finishing touch of the furniture.

Hunched rails and clamp chips are often butted together between legs below the table top to reinforce the structure and divide the space. This square table's floral clamp chips and hunched rails absolutely distinguish the furniture with a rare elegant delicacy.

052

攒牙板瓜棱腿方桌

Square Table with Assembled Apron Plates and Legs of Melon-shaped Sections

89×89×83cm

方桌由花梨木制成，花梨木色泽淡雅，纹理流畅，有文人气息。

桌面四面攒边，边抹以格角榫连接，面心平装，边抹做冰盘沿压边线处理。桌面下接四腿，腿与桌面高低榫连接。四腿出挓明显，以增加稳定性。腿截面为瓜棱状，即四面打洼起阳线委角，线脚变化丰富。

桌面与腿之间采用攒牙板做法，用横长竖短的直材攒接出海棠形的开光，牙头则用横短竖长的直材接出海棠形的开光，牙头边缘一角做委角处理。如此，形成牙条三横向海棠开光、牙头一竖向海棠开光的造型。此攒牙板做法为罗锅枨和矮老搭配的变体，两罗锅枨和矮老横竖材重新搭配组合而成。海棠开光边缘和牙板边缘皆起阳线，勾勒轮廓，增加细节。

此件方桌特殊的攒牙板做法，和瓜棱式腿的结合，使得整件桌子线脚丰富，精巧雅致。

见《故宫博物院藏文物珍品大系・明清家具》（上）79页。

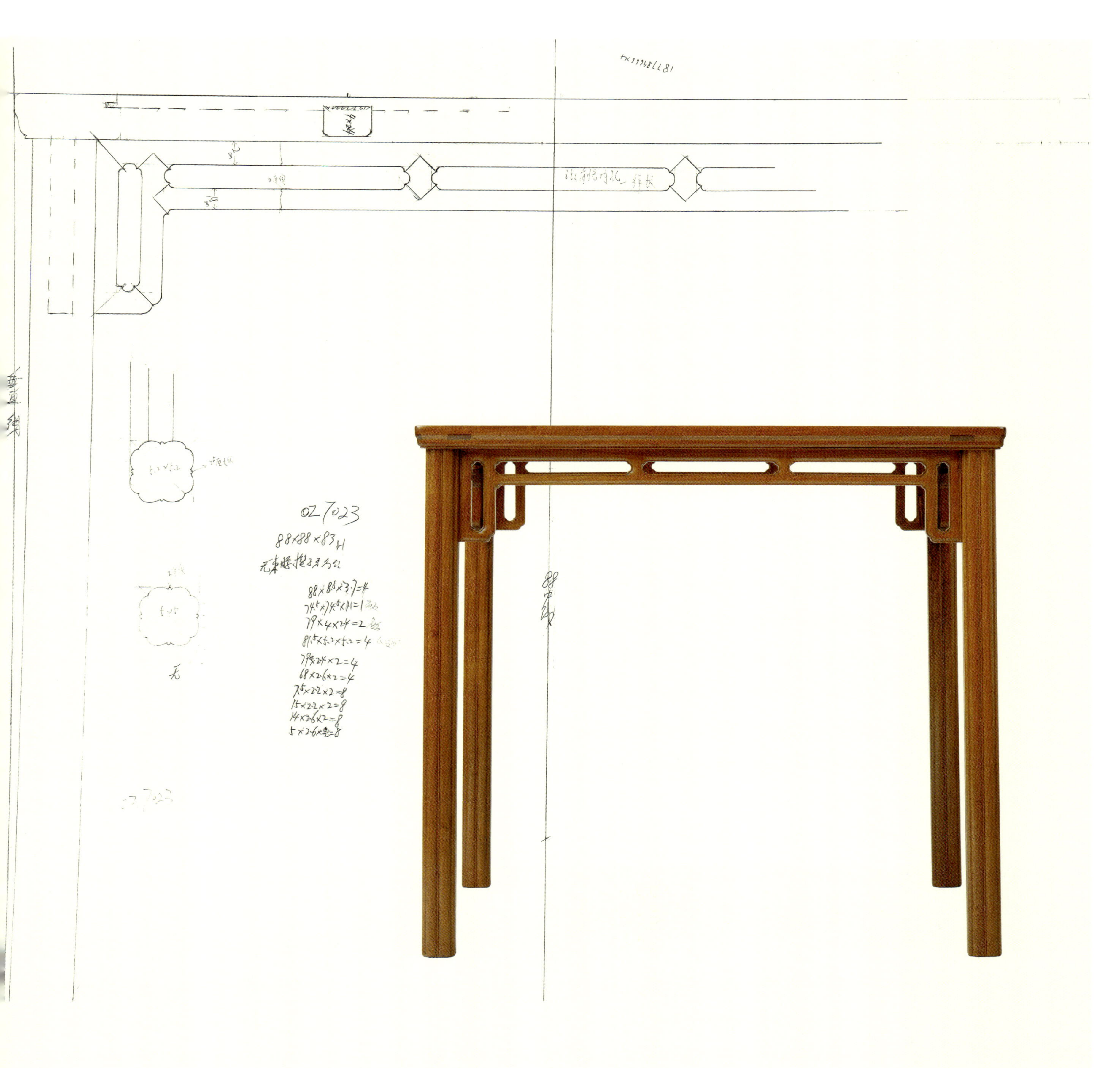

The square table is made of *Huali* wood of selected refined material with elegant colour, smooth and natural veining, exhibiting aesthetics of literati.

Table top is constructed with assembled frames by mitered corner bridle joints, and embedded flat central panel, with edges narrowed down widthways (also called ice-plate edge) and cut in a rabbet along the bottom rim as the borderline; below the table top connects four splayed legs by uneven double tenons to maintain the steadiness of the structure; grooves are chiseled out lengthways along legs with an embossed line carved in each trough line to have melon-shaped sections of four legs taken shape.

Apron plates between legs are assembled with straight sticks into three horizontal begonia-shaped traceries on each side; corner plates of vertical begonia-shaped tracery have their tips dented to keep a unified decorative style. Such assembled apron plate is a variant of the combination of hunched rails and short rods, a re-collocation of two horizontal hunched rails and vertical short rods; all the apron plates are outlined with embossed borderlines to stress the silhouette and add details to the furniture.

The unique assembled apron plates combined with legs of melon-shaped sections vigorously add much detail to the square table, make it more of an elegant and exquisite artwork.

Reference to *Compendium of Collections in the Palace Museum: Furniture of Ming and Qing Dynasties*, volume I, page 79.

053

一腿三牙罗锅枨方桌

Square Table with Hunched Rails
and Three-plate Butted Legs

86×86×83cm

方桌由花梨木制成，花梨木色泽温润，纹理优美。

桌正方，桌面四面攒边，面心平装，边抹做冰盘沿，下再贴一层垛边起阳线，挡住腿上安装的部分牙条和角牙，造成桌面边抹是由原有边抹加垛边组成的假象，视觉上增加了边抹的厚度。

桌面之下安四圆腿，腿与桌面以高低榫连接。四腿八挓明显，形成下大上小的稳定感。

两腿之间安刀板牙子和高拱罗锅枨，刀板牙子起阳线，罗锅枨素混面，拱起部分直接顶到刀板牙子牙条之上。桌面四角之下，以45度角位置安小角牙，与圆腿连接。如此，圆腿前面、侧面和45度角位置皆安牙子，形成一腿三牙。

此方桌造型简素，一腿三牙的处理方法是方桌的一种基本制式。

见王世襄《明式家具研究》乙49，页99。

The square table is made of *Huali* wood of selected refined material with elegant colour, smooth and natural veining.

Square table top is constructed with assembled frames and embedded flat central panel, with edges narrowed down widthways (also called ice-plate edge) and attached with stacked edges cut in a rabbet along the bottom rim as the borderline to block partial apron plates installed between legs below the top panel, so as to exhibit a false appearance and increase the thickness of table top's edges visually.

Below the table top installs four splayed legs with round sections by uneven double tenons to maintain the steadiness of the structure.

Unadorned straight apron plates are butted between legs by T-bridle joints (clamp joint) and outlined with embossed borderlines; hunched rails with cambered surface are polished and installed below apron plates, with each raised part touching the bottom rim of the plate. Below four table top corners connects extra corner plates at 45-degree angle, which are also butted into legs vertically; thus, each leg is butted with three plates on three sides.

This square table is of clean and simple structure; and the three-plate butted leg is a basic structural form in ancient Chinese furniture.

Reference to *Connoisseurship of Chinese Furniture: Ming and Early Qing Dynasties* by Wang Shixiang, B49, page 99.

054

一腿三牙罗锅枨方桌

Square Table with Hunched Rails and Three-plate Butted Legs

89×89×83cm

方桌由花梨木制成，花梨木色泽淡雅莹润，纹理细腻流畅。

桌正方，桌面四面攒边，面心平装，边抹做冰盘沿压边线线脚处理。桌面之下再接一层垛边，垛边边缘亦起阳线。桌面下增加一层垛边，造成桌面边抹是由原有边抹加垛边组成的假象，视觉上增加了边抹的厚度。

桌面四角下接四腿，圆腿四面八挓，挓度较大，形成稳定感。桌面下两腿之间安刀板牙子和高拱罗锅枨。刀板牙子牙条部分坠下少许，与罗锅枨拱起部分连接，形成整体。刀板牙子牙头部分略高起，与罗锅枨罗锅曲线部分围合出S形空间，增加通透性。桌面四角下、圆腿外，又安小角牙，形成了腿上端三个方向都做牙子的形式，是为一腿三牙。

此方桌造型简洁大气，雍容典雅。

见王世襄《明式家具研究》乙49，页99。

The square table is made of *Huali* wood of selected refined material with elegant colour, smooth and natural veining.

Square table top is constructed with assembled frames and embedded flat central panel, with edges narrowed down widthways (also called ice-plate edge), cut in a rabbet along the bottom rim as the borderline, then attached with stacked edges outlined with embossed borderlines, so as to exhibit a false appearance and increase the thickness of table top's edges visually.

Below the table top installs four splayed legs with round sections to maintain the steadiness of the structure. Unadorned straight apron plates are butted between legs by T-bridle joints (clamp joint) and outlined with embossed borderlines; the horizontal part of apron plate on each side drops a bit to connect the below hunched rail to serve as one component; raised curve of corner plates and the hunched silhouette of rail together constitute an s-shaped openwork at two upper corners on each side to add a sense of transparency; below four corners of top panel installs four extra corner plates at 45 degree angle, which are also butted into legs vertically, to form the structure of three-plate butted legs.

This square table is of clean and simple structure, an ethereal beauty of elegance and magnificence.

Reference to *Connoisseurship of Chinese Furniture: Ming and Early Qing Dynasties* by Wang Shixiang, B49, page 99.

055

一腿三牙灵芝卡子花罗锅枨方桌

Square Table with Hunched Rails, *Ruyi*-shaped Cloud Chips and Three-plate Butted Legs

89×89×83cm

方桌由花梨木制成，花梨木色泽淡雅莹润，纹理细腻流畅。

桌正方，桌面四面攒边，面心平装，边抹做冰盘沿压边线线脚处理。桌面下再围垛边一圈，垛边边缘亦起阳线。

桌面下接瓜棱腿，四腿八挓，收分明显，形成下大上小的稳定感。桌面下两腿间装花牙子和罗锅枨。花牙子牙条平素，牙头翻转出卷草，边缘起阳线，以勾勒轮廓。罗锅枨高拱，罗锅曲线做两次委角处理，边缘起阳线。花牙子和罗锅枨之间安如意云头形灵芝纹卡子花两朵，均匀分布其间。桌面四角下、瓜棱腿外，再安卷草纹小角牙，形成一腿三牙。花牙子、罗锅枨、卡子花、角牙，各自翻转曲线，形成具有丰富线脚变化的错落空间，精致秀雅。

见王世襄《明式家具研究》乙50，页100。

The square table is made of *Huali* wood of selected refined material with elegant colour, smooth and natural veining.

Square table top is constructed with assembled frames and embedded flat central panel, with edges narrowed down widthways (also called ice-plate edge), cut in a rabbet along the bottom rim as the borderline, then attached with stacked edges outlined with embossed borderlines.

Below the table top connects four splayed legs of melon-shaped sections to maintain the steadiness of the structure; between legs installs apron plates and hunched rails; unadorned straight plates outlined with embossed borderlines run horizontally and roll up as floral leaves at upper corners; hunched rails outlined with embossed borderlines are dented twice at turning points; between apron plate and hunched rail installs two evenly distributed *Ruyi*-shaped cloud chips on each side; below four corners of top panel installs four extra corner plates with embossed floral borderlines at 45 degree angle, which are also butted into legs vertically, to form the structure of three-plate butted legs. Apron plates, hunched rails, floral chips, corner plates are respectively decorated by graceful curves and exquisite carvings to create a space of variations.

Reference to *Connoisseurship of Chinese Furniture: Ming and Early Qing Dynasties* by Wang Shixiang, B50, page 100.

056

罗锅枨矮老马蹄腿方桌

Square Table with Hunched Rails, Short Rods and Hoof Feet

90×90×81cm

方桌由黄花梨制成，精选黄花梨好料，纹理流畅自然，色泽莹润，油性充足。

桌正方，桌面四面攒边，面心平装，边抹做冰盘沿压边线线脚处理。

桌面下接束腰，束腰矮素无饰。束腰下接四腿，四腿与桌面以抱肩榫连接。腿竖直方正，自束腰顺势而下，于末端以内翻马蹄收尾。束腰下两腿间安素直牙条，牙条与四腿边缘起阳线，以勾勒轮廓。两腿之间安罗锅枨加矮老。罗锅枨拱起曲线劲挺有力，罗锅枨之上安两矮老。

此方桌为典型的束腰式结构，采用罗锅枨加矮老搭配，使得桌面下空间通透空灵。

It's made of yellow *Huali (Huanghuali)* wood of highly refined material with smooth and natural veining, elegant colour and rich oiliness.

Square table top is constructed with assembled frames and embedded flat central panel, with edges narrowed down widthways (also called ice-plate edge) and cut in a rabbet along the bottom rim as the borderline.

Below the top panel installs an unadorned girdled waist and four legs by vertical mitered corner bridle joints; legs of square sections outlined with embossed borderlines run straightly downward and tilt inward a bit at end as hoof feet; between legs connects a straight apron plate outlined by embossed borderlines, a hunched rail with clean curves and two short rods in between on each side.

This square table is built in typical girdled waist structure, with hunched rails and short rods, to create an ample space under the top panel.

清早期　佚名《清宫珍宝皕美图》册

Early Qing Dynasty (1644-1911)
Illustration of *Two Hundred Treasures from Royal Collection of Qing Dynasty* (album)
By Anonymity

057

螭龙纹三弯腿方桌

Square Table with Bas-relief of Dragon Patterns and Three-curved Legs (Cabriole Legs)

88×88×81cm

方桌由黄花梨制成，黄花梨色泽温润，纹理流畅，油性充足。

桌正方，桌面四面攒边，面心平装，边抹做冰盘沿压边线线脚处理。桌面下收束腰，束腰较矮。束腰下接三弯腿，腿与桌面抱肩榫连接。三弯腿曲线兜转有力，至末端外翻出卷草收尾。束腰下两腿间安壸门牙条，牙条上浅浮雕螭龙卷草纹饰，两行龙对首而行，中间为壸门花心向上翻卷出的卷草灵芝纹饰。牙条和腿连接处浅浮雕兽面纹饰，又像是模仿铜饰角的做法。

S形螭龙纹角牙安于牙条和腿之间，螭龙尾部翻处卷草，优雅自然。

此方桌造型别致，腿足、牙条和角牙多加雕饰，尽显雍容静雅。

It's made of yellow *Huali (Huanghuali)* wood with elegant colour, smooth veining and rich oiliness.

Square table top is constructed with assembled frames and embedded flat central panel, with edges narrowed down widthways (also called ice-plate edge) and cut in a rabbet along the bottom rim as the borderline; below the top panel installs a short girdled waist and four three-curved legs (cabriole legs) by vertical mitered corner bridle joints; legs curve dramatically, tilt outward a bit and turn to rolled-up leaf feet at end; below the girdled waist installs *Kun*-gate apron plates carved with bas-relief dragons and arabesque patterns (two dragons flying toward a ganoderma-shaped floral pattern in the middle on each side); at joints of plates and legs is carved with abstract beast-face motif resembling copper corner protector on bronze ware; corner plates are carved into three-dimensional s-shaped dragons with floral tails and butted between apron plates and legs.

This square table is of exquisite form, with legs, feet, apron plates and corner plates heavily decorated by carvings and bas-relief to fully demonstrate its extravagance and elegance.

058

四面平马蹄腿条桌

Narrow Table with Flat-corner Structure (Corners Butted with Mitered Corner Bridle Joints) and Hoof Feet

104×48×82cm

桌长方，黄花梨制成。四面平式，桌面四面攒成，大边和边抹以格角榫连接。面心平装薄板，下以穿带支撑。桌面下紧挨着大边安长直牙条，与大边齐平，牙条两侧做圆角曲线与腿连接。牙条光素无饰，甚至不起阳线，不做线脚处理，任其光素。

四腿接于桌面四角之下，腿截面方形，略起挓，竖直向下，渐渐收细，于末端内翻马蹄收尾，马蹄低矮。

此件长方桌采用四面平和内翻马蹄腿的搭配，塑造出修长、简洁的造型，极尽简致，充分体现了“少即是多”的设计理念。

参考清华大学艺术博物馆所藏黄花梨四面平马蹄腿长方桌。

This rectangular table is made of yellow *Huali (Huanghuali)* wood, with flat-corner structure (corners butted with mitered corner bridle joints); the table top is constructed with assembled frames butted together by mitered corner bridle joints and embedded flat central panel, and is strained by sliding dovetail splines at the bottom. Below the long edges of the table top directly installs straight apron plates with smooth curves at corner joints; unadorned plates are left without any embellishment.

Below table top corners connects four splayed legs with square sections, tapering lengthways and rolling up inward as low hoof feet at end.

This rectangular table is built with flat-corner structure combined with low hoof feet to construct a slender and clean structure of extreme simplicity, and to fully demonstrate the aesthetics of "less is more".

Reference to Yellow *Huali* Rectangular Table with Flat-corner Structure and Hoof Feet from Tsinghua University Art Museum.

明万历
《元曲选图》插图

Wanli Period of Ming Dynasty (1573-1620)
Illustration of *Selected Illustrations of Verses of the Yuan Dynasty*

059

高束腰霸王枨马蹄腿条桌

Narrow Table with High Girdled Waist, Curved Diagonal Rails and Hoof Feet

153×48×82cm

条桌由紫檀制成，紫檀色泽沉穆稳重，纹理隐现表面。

条桌长方，桌面四面攒边，面心平装，边抹做冰盘沿线脚处理。桌面下承束腰，束腰四角安短柱，短柱之间装平板。束腰略高，厚度高于桌面和牙条厚度。束腰之下承四腿，腿截面方形，自束腰竖直而下，逐渐收缩变细，于末端内翻马蹄收尾，马蹄兜转有力。腿偏上位置安霸王枨，霸王枨三弯曲线，下端与腿勾挂榫连接，上端与桌面下的穿带连接。

此条桌造型简洁，使用桌面、束腰、内翻马蹄腿、霸王枨四种构件组成，结构坚固，造型简洁，结构与造型完美结合。条桌体形修长，气质俊逸。

明崇祯
《新撰醋葫芦》插图
笔耕山房刊本

Chongzhen Period of Ming Dynasty (1628-1644)
Illustration of *Vinegar Gourd* (new edition), Bi Geng Shan Fang's edition

The narrow table is made of red sandalwood with dark and solemn colour and subtle veining.

The rectangular table top is constructed with assembled frames and embedded flat central panel, with edges narrowed down widthways (also called ice-plate edge) and cut in a rabbet along the bottom rim as the borderline; below the top panel installs a high girdled waist assembled by four short rods of square sections at corners and four straight plates in between; and the girdled waist is a bit higher than the thickness of top panel and apron plates; below the waist installs four legs of square sections running straightly downward, tapering lengthways and rolling up inwardly as hoof feet at end; at upper parts of legs connects three-curved diagonal rails by wedged crook joints; and the other ends of the four diagonal rails are butted into sliding dovetail splines at the bottom of table top.

This table is of simple and clean structure, built with four components of table top, girdled waist, inward hoof feet and curved diagonal rails to reinforce the steadiness and maintain a simple appearance, as a perfect combination of form and structure. Its slender silhouette exhibits a beauty of elegance.

060

高束腰霸王枨马蹄腿翘头桌

Narrow Table with Raised Short Edges, High Girdled Waist, Curved Diagonal Rails and Hoof Feet

117.5×41.5×79.5cm

此桌由黄花梨制成，选自黄花梨大料原材，纹理流畅大气，异于小料的多变花纹。

桌面精选黄花梨独板，做冰盘沿线脚。独板短边处拍抹头做翘头，既遮挡了断面不美的纹理，又可以减少独板变形，还塑造出精致上扬的翘头，以勾勒轮廓。

桌面下安束腰，束腰四角的短柱为四腿向上延伸而成，短柱之间装平板，束腰正中设抽屉。束腰之下装四腿，腿修长挺直，略出挓，于末端内翻马蹄收尾。束腰下两腿间安长直牙条，牙条两端与腿连接。四腿上端偏上位置出霸王枨，霸王枨S形曲线，下与腿以钩挂榫连接，上与桌面下穿带连接。

整件条桌体形修长，比例协调，造型简练，气质独特。

明　抱瓮老人《今古奇观》插图
明末吴郡宝翰楼刊本

Ming Dynasty (1368-1644)
Illustration of *Anecdotes of the Past and the Present*, Baohanlou's (at County Wu of late Ming Dynasty) edition
By Oldman of Baoweng

It's made of one-piece yellow *Huali (Huanghuali)* wood of highly refined material with smooth and natural veining different from changeable veining of small materials.

The table top is made of one-piece *Huali (Huanghuali)* wood of highly refined material, with edges narrowed down widthways (also called ice-plate edge) and cut in a rabbet along the bottom rim as the borderline; its two raised short edges can hide the unattractive veining of sections, reduce the panel's deformation, and also build up an exquisite silhouette.

Below the top panel installs a girdled waist with its four corners butted with long components of square sections which extend through apron plates as four legs; unadorned straight plates are embedded between waist's corner components; and a drawer is built in the middle of the girdled waist; four splayed legs run straightly downward and tilt inwardly as hoof feet at end; between legs connects straight apron plates; at upper parts of legs installs s-shaped curved diagonal rails by wedged crook joints; and the other ends of the four diagonal rails are butted into sliding dovetail splines at the bottom of table top.

This narrow table built in proper scale, is of slim and clean structure, exhibits a unique beauty of simplicity and elegance.

061

罗锅枨马蹄腿条桌

Narrow Table with Hunched Rails and Hoof Feet

153×76×81cm

条桌由花梨木制成，花梨木色泽淡雅温润，纹理细腻流畅，文气十足。

桌长方，桌面四面攒边，面心平装，边抹做冰盘沿压边线线脚处理。桌面下收束腰，束腰光素。束腰下接四腿，腿修长挺劲，竖直而下，逐渐收细，与末端内翻马蹄结尾，马蹄兜转有力。四腿出挓，收分明显，以增加稳定性。束腰下、两腿间安长牙条，牙条修长光素，牙条两侧向下出弧线，与四腿连接。牙条和腿边缘皆起阳线，一直贯穿至腿足末端之马蹄。

牙条之下、两腿之间安罗锅枨，做素混面线脚，混面曲线圆润饱满。罗锅枨高拱曲线微起，精巧有力。

此条桌造型精炼，工艺精湛，气质独特。

清　殷奇《春宫图》册　美国波士顿美术馆

Qing Dynasty (1644-1911)
Illustration of *Erotica Stories* (album)
By Yin Qi
Museum of Fine Arts, Boston

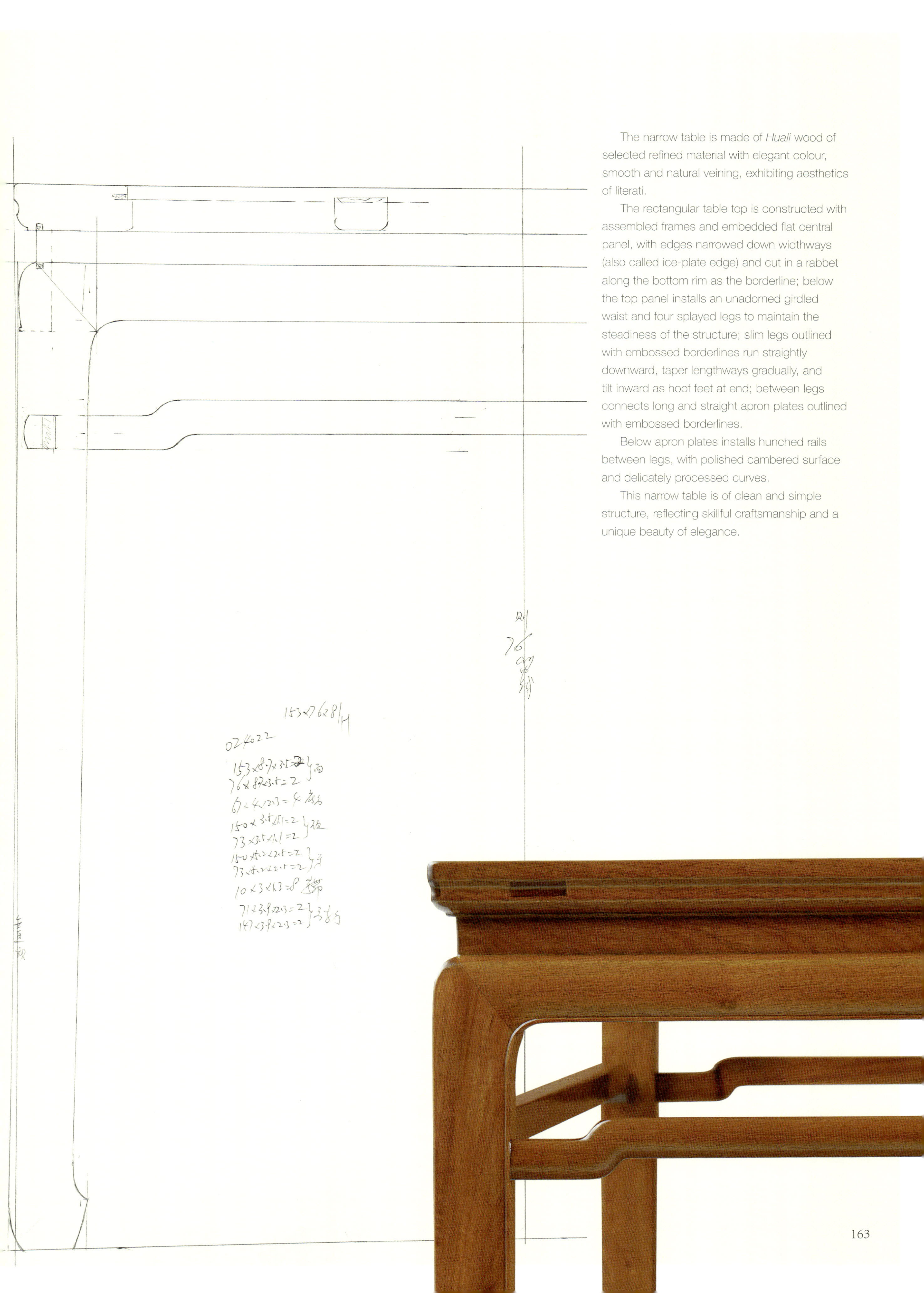

The narrow table is made of *Huali* wood of selected refined material with elegant colour, smooth and natural veining, exhibiting aesthetics of literati.

The rectangular table top is constructed with assembled frames and embedded flat central panel, with edges narrowed down widthways (also called ice-plate edge) and cut in a rabbet along the bottom rim as the borderline; below the top panel installs an unadorned girdled waist and four splayed legs to maintain the steadiness of the structure; slim legs outlined with embossed borderlines run straightly downward, taper lengthways gradually, and tilt inward as hoof feet at end; between legs connects long and straight apron plates outlined with embossed borderlines.

Below apron plates installs hunched rails between legs, with polished cambered surface and delicately processed curves.

This narrow table is of clean and simple structure, reflecting skillful craftsmanship and a unique beauty of elegance.

062

蕉叶纹高束腰马蹄腿条桌

Narrow Table with High Girdled Waist Carved with Banana-leaf Motif and Hoof Feet

175.5×47.5×86cm

条桌由紫檀制成，精选紫檀好料，纹理细腻，牛毛纹清晰可见，棕眼满密，油性充足。

桌长方，桌面四面攒边，面心平装，边抹做冰盘沿压边线线脚处理，冰盘沿之上起阳线，围合成圆角长方形。桌面之下接高束腰，束腰打洼处理，打洼表面浅浮雕蕉叶纹。束腰之下安托腮，托腮做混面压边线线脚处理，混面饱满。

束腰之下接长牙条和四腿。腿截面方正，直腿竖直而下，下端内翻马蹄，内翻马蹄表面阴刻拐子纹饰。马蹄之下又一矮台，上卷珠结尾。牙条长直，中间下坠出一层，向上翻出卷珠纹。牙条和四腿连接之处做云头卷珠纹饰，自卷珠和边缘各起一阳线，横向连接至牙条中心下坠的卷珠处，竖向连接至腿足下端内翻马蹄。

此条桌造型简练，在细节之上做适当雕刻装饰，雕工细腻精到。整器实为精工之作，为典型的清式家具。

The narrow table is made of red sandalwood of selected material with refined and smooth veining, visible ox-hair and pinhole veining (traces of capillary tubes of the plant), and rich oiliness.

The rectangular table top is constructed with assembled frames and embedded flat central panel, with edges narrowed down widthways (also called ice-plate edge) and cut in a rabbet along the bottom rim as the borderline; frames of top panel are outlined with embossed borderlines to compose a rectangular with round corners; below the table top installs a high girdled waist gouged out a round low-lying groove carved with bas-relief of banana-leaf motif; below the waist connects moulding polished in cambered surface and cut in rabbets as borderlines.

Straight apron plates and four legs of square sections are butted below the moulding; legs run straightly downward and roll up inwardly as hoof feet with intaglio of fretwork at end; small foot pad is butted below each foot and carved with a rolled-up pearl; long and straight apron plates are also carved with rolled-up pearls in the middle and cloud patterns at corner joints; apart from the cloud motif at each corner, apron plates and legs are all outlined with embossed borderlines.

This narrow table is of clean and simple structure, with detailed and exquisite carvings exactly highlighting the advantage of the red sandalwood's texture, demonstrating refined craftsmanship of ancient woodworkers and representing the typical furniture of Qing Dynasty.

063

夔龙纹马蹄腿条桌

Narrow Table with Bas-relief of Dragon Patterns and Hoof Feet

74×35×80cm

桌长方，紫檀制，桌面四面攒边，面心平装，边抹做打洼混面压边线线脚处理。

桌面下安束腰，束腰较高，束腰之上浅浮雕三组纹饰，正中一组为西洋卷草纹，两侧为拐子纹，束腰转角处再雕一半西洋卷草纹，侧面雕另一半。

束腰之下安托腮，托腮做冰盘沿压边线线脚处理。托腮之下安牙条和四腿。四腿方正竖直，略起挓度，于末端内翻拐子纹，再在桌腿外侧浅浮雕卷草纹，向上扬起，优雅自然。牙条素直，牙条和腿足边缘皆起宽皮条线加洼儿，直腿足末端收于内翻拐子纹之上。

牙条之下、方腿之间装花牙子，牙条正中雕蝙蝠纹，两侧浅浮雕和透雕结合雕夔龙纹，一直延续至两侧牙头，形成华丽的雕刻。

此条桌在简洁精致的造型上，采用浅浮雕、透雕相结合的方法修饰细节，形成了浓华绚丽的雕刻装饰风格，气质华贵典雅。

The rectangular table top is constructed with assembled frames and embedded flat central panel, with edges gouged out a round low-lying groove and cut in a rabbet along the bottom rim as the borderline.

Below the table top installs a high girdled waist carved with three groups of bas-relief including arabesque pattern in the middle, fretwork pattern on two sides, arabesque pattern at each corner.

Below the girdled waist installs the moulding narrowed down widthways (also called ice-plate edge) and cut in a rabbet along the bottom rim as the borderline, apron plates and four splayed legs of square sections; legs roll up inwardly as fretwork feet at end and carved with elegant bas-relief of floral pattern crawling up on exteriors of lower parts; unadorned apron plates and legs are outlined with embossed belt-lines (wider than the usual borderlines) gouged out a round low-lying groove in the middle.

Between legs on two sides of the narrow table connects luxuriously decorative plates carved with delicate bat motif in the middle, bas-relief and openwork of abstract dragon design as the background.

This narrow table exhibit an obvious elegant opulence by adopting various carving skills of bas-relief and openwork to embellish details of the furniture.

064

梅花纹马蹄腿条桌

Narrow Table with Bas-relief of Plum Blossoms and Hoof Feet

155×45×88cm

条桌由紫檀制成，紫檀色彩沉穆稳重，纹理细腻，正适合满雕纹饰。

桌长方，桌面西面攒边，面心平装。桌面下接束腰，束腰略向内收，与桌面分出层次。束腰之下安四腿和牙条。四腿方直，竖直而下，于末端内翻马蹄收尾。四腿略起挓，形成下大上小的稳定感。牙条平直，侧端转角与腿连接。牙条与腿足连接处装角牙，角牙为透雕折枝梅纹，枝干遒劲，梅朵怒放，生动传神。

桌面边抹立面、束腰、牙条和腿足之上满雕梅枝梅花纹饰，枝干疏影横斜，花苞欲放，梅朵争艳，热闹非凡。

此条桌在束腰、内翻马蹄造型基础之上，在表面满雕梅花纹饰，与光素平直的桌面形成鲜明对比，艺术风格独特。

The narrow table is made of red sandalwood with dark and solemn colour, refined and smooth veining, hence an ideal material for carving and sculpting.

The rectangular table top is constructed with assembled frames and embedded flat central panel; below the table top installs a girdled waist to tighten the silhouette of the edge; four splayed legs and apron plates are butted below the waist to maintain the steadiness of the structure; legs of square sections run straightly downward and roll up inwardly as hoof feet at end; corner plates are butted at joints of apron plates and legs, and carved with quite realistic openwork of plum blossoms stretching branches and twigs, bursting into full bloom, exuberating and crawling across edges, girdled waist, apron plates and legs.

This narrow table is built with basic structures of girdled waist and hoof feet, and fully covered by carvings of plum blossoms to present a sharp contrast to the unadorned flat table top, exhibiting a unique aesthetic style.

065

四面平马蹄腿条桌

Narrow Table with Flat-corner Structure (Corners Butted with Mitered Corner Bridle Joints) and Hoof Feet

211×48×85cm

条桌为铁梨木老料制成，色泽极为沉稳，稍加擦拭，即光泽莹润，出现包浆。

明末清初名士张岱的《陶庵梦忆》中有这样一段记载：“癸卯，道淮上，有铁梨木天然几，长丈六，阔三尺，滑泽坚润，非常理。淮抚李三才百五十金不能得，仲叔以二百金得之，解维遽去。淮抚大恚怒，差兵蹑之，不及而返。”一件五米多长的铁梨木天然几，所费靡多也就罢了，竟然惹得淮抚派兵追赶，急赤白脸的样子跃然纸上，也可见当时人对于铁梨木家具的看重。本例铁梨木条桌，以厚独板为面，还保留铁梨木老料本身的皮壳，古家具领域多将这种独板面的做法称为“一块玉”，可见其珍。

既有上好的桌面，其他部分的设计自然也迎刃而解，采取四面平结构，植以四足，连以牙板，任其光素，不加任何雕饰甚至线脚，全在于各部分比例的拿捏。牙板两端饰一块宽料挖成的大弧度委角，与腿足柔和交圈，并与下方兜转的马蹄腿相呼应，整体形成外方内圆的审美情趣。

The drawing table is made of old lignumvitae covered by a thin layer of patina, presenting dark and solemn colour, elegant and subtle gloss.

According to the anthology *Dream Recollections of Taoan* written by Zhang Dai (self-entitled as Taoan, famous scholar in the late Ming Dynasty and early Qing Dynasty, 1597-1689), "in the year of 1663, a narrow table made of natural lignumvitae appeared; its smooth texture and elegant gloss are of rare and unique beauty. Highly ranked official Li Sancai, the governor of grain transportation, wanted to buy it with 150 gold; however, my uncle acquired it with 200 gold and then carried it away instantly. The governor was in fury and ordered soldiers to pursue; but they lost him anyway in the end." A five-metre-long narrow table made of natural lignumvitae, not only could cost a fortune, also drew the attention of highly ranked official and forced him to employ military power to get it; the story easily reflects people's affection for lignumvitae furniture in the early Qing Dynasty (1644-1911). This narrow table is built with a thick central panel made of old one-piece lignumvitae covered by a thin layer of patina, which is also called "a block of jade" to emphasize its rarity and valuableness among collectors.

Aside from the highly selected table top, the drawing table is constructed with flat-corner structure (corners butted with mitered corner bridle joints, also called cubic-cone rice-pudding joint); below the top panel installs unadorned apron plates and four legs of proper scale; at corner joints, straight turnings of apron plates are processed into curves of large radius to connect four legs and echo to the curves of hoof feet at bottom, exhibiting the aesthetic philosophy of "square externally and round (sophisticated) internally"(not only referring to the appearance, but also the guideline of personal conduct).

066

裹腿罗锅枨画桌

Drawing Table with Hunched Rails Bound to Legs
(Puttee-styled Hunched Rails)

195×80×82cm

画桌由紫檀制成，紫檀色泽深沉，纹理隐现，棕眼浓密。

桌长方，桌面四面攒边，大边与抹头用格角榫连接。面心平装薄板，薄板下安穿带，与大边榫接。边抹做素混面线脚处理，混面圆润，与下面的裹腿造型相呼应。

桌面下四角安四腿，腿与桌面高低榫连接。腿截面为圆形，四腿八挓，收分明显，以增加稳定感。桌面下、两腿间安罗锅枨，罗锅枨高拱部分直接顶到桌面下。罗锅枨高出腿面，与侧面罗锅枨交圈，形成裹腿枨结构。裹腿枨式罗锅枨连接桌面和腿足，既是加固结构，又是造型元素。

此画案模仿竹制家具的做法，做裹腿枨式罗锅枨，且以素混面边抹相呼应，造型简致静雅。

The drawing table is made of red sandalwood with dark and solemn colour, subtle and smooth veining, visible ox-hair and pinhole veining (traces of capillary tubes of the plant).

This rectangular table top is constructed with assembled frames butted together by mitered corner bridle joints and embedded flat central panel, and is strained by sliding dovetail splines butted into long edges at the bottom; unadorned cambered edges run smoothly to echo the puttee-styled rails below.

Below the table top installs four splayed legs of round sections by uneven double tenons to maintain the steadiness of the structure; between legs connects hunched rails with its raised parts touching the bottom rims of the top panel; hunched rails are carved as puttee-styled rails, wrapping four legs on the exterior, connecting the table top and legs, so as to reinforce the steadiness and also to highlight its unique form.

This drawing table is built in a style to imitate bamboo furniture's craftsmanship through adopting puttee-styled hunched rails to echo its cambered edges of the top panel, exhibiting an aesthetics of simplicity and serenity.

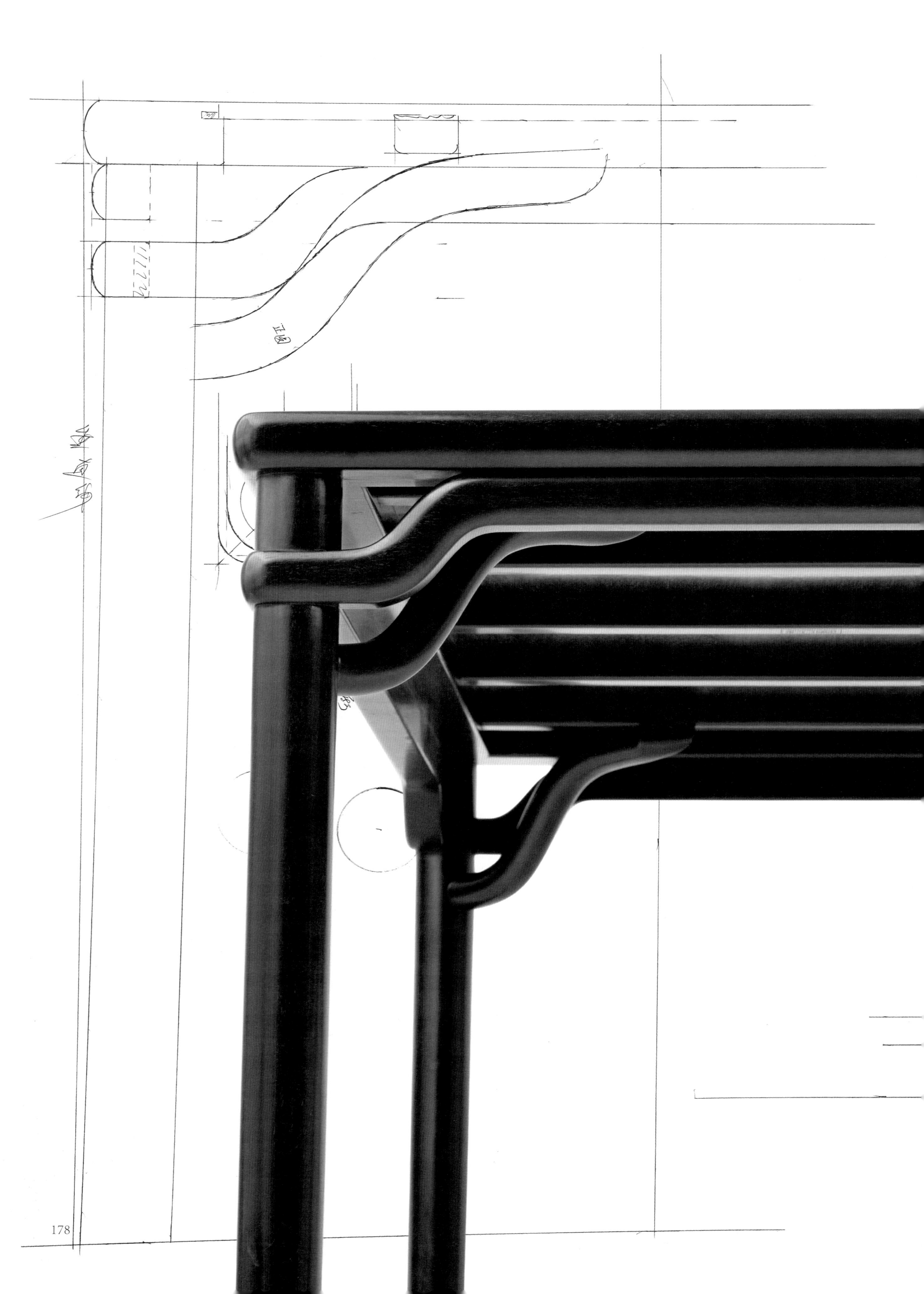

清早期　佚名《清宫珍宝皕美图》册

Early Qing Dynasty (1644-1911)
Illustration of *Two Hundred Treasures from Royal Collection of Qing Dynasty* (album)
By Anonymity

067

裹腿罗锅枨画桌

Drawing Table with Hunched Rails Bound to Legs
(Puttee-styled Hunched Rails)

196×88×83cm

此画桌由黄花梨制成，黄花梨精选大材入料，纹理流畅大气，麦穗纹密集，色泽淡雅莹润，油性充足。

画桌长方，桌面四面攒边，大边与抹头以格角榫连接，面心平装薄板，薄板下以穿带连接，并与两大边榫接。边抹做素混面，上阔下敛。

桌面下安四圆腿，圆腿粗硕，腿与桌面高低榫连接，四腿八挓，收分明显，以增加稳定感。桌面下、两腿间安裹腿直枨，直枨做劈料，即在枨上做两个平行混面，混面饱满圆润。裹腿枨与圆腿连接处下缘，再安相同混面牙条一段，在桌角下造成三段素混面并置的效果。

裹腿枨之下，再安罗锅枨，罗锅枨高出腿面，与侧面罗锅枨交圈，也呈现裹腿枨效果。圆腿上部再安霸王枨，霸王枨三弯，下端与圆腿以勾挂榫连接，上端与桌面下穿带连接。

此画桌重复使用裹腿枨做法，在桌角下、腿上部造成多层素混面并置的效果，是典型的仿竹家具制式。

This drawing table is made of yellow *Huali (Huanghuali)* wood of refined large material with smooth and dense wheat veining, elegant colour, and rich oiliness.

This rectangular table top is constructed with assembled frames butted together by mitered corner bridle joints and embedded flat central panel, and is strained by sliding dovetail splines butted into long edges at the bottom; unadorned cambered edges are narrowed down widthways.

Below the table top installs four splayed legs of round sections by uneven double tenons to maintain the steadiness of the structure; below the top panel installs puttee-styled straight rails split into two paralleled cambered edges; below the corner joints, a short extra corner plate with same style is butted to create an illusion of juxtaposition of three paralleled cambered edges.

Below the puttee-styled straight rails installs hunched rails which are also processed to wrap four legs on the exterior, with raised parts touching the bottom rims of the straight rails above to create an illusion of juxtaposition of three paralleled cambered edges; four three-curved diagonal rails are butted into upper parts of legs by wedged crook joints on bottom end, and then fixed into sliding dovetail splines at the bottom of table top on the other end.

This drawing table is featured with the repetition of puttee-styled rails to create a unique visual effect of juxtaposition of multiple paralleled cambered edges at corner joints and upper parts of legs, which is a typical style for bamboo-inspired furniture.

068

螭龙捧寿纹带底座供桌

Altar Table Carved with Openwork of Dragons Holding up Chinese Character "寿" (Longevity) (with a Bottom Base)

81×44.5×83cm

桌由白酸枝制成，面长方，四面攒边，以格角榫连接，面心平装硬板。边抹做冰盘沿压边线线脚。桌面下安高束腰，四腿穿过托腮，向上延伸出短柱，做束腰边缘。束腰正中再安矮老式短柱，装两绦环板。绦环板上开海棠形开光，以阳线勾勒边缘。开光内透雕双龙捧寿纹。

束腰之下装托腮，托腮之下安四腿和壶门牙条。四腿三弯，先向外膨出，再向内收敛，于末端向外翻转出卷草纹，曲线弯折有力。牙条上壶门曲线流畅自然，与三弯腿自然衔接。牙条与三弯腿连接之处，浅浮雕云头纹饰，似镶嵌在家具转角处的铜包角。

四腿之下安台座式底座，台座四面攒边，中平装硬板，边抹做混面压双边线线脚处理，混面饱满圆润。台座下安高束腰，高度比桌面下束腰略高，以分出主次。束腰亦以三短柱界出两段空间，镶绦环板，绦环板上开海棠形和自然形三段开光，海棠形开光内透雕卷草纹饰。束腰下安壶门牙条，牙条边缘起阳线。

此长方桌结构谨严，桌面和台座处理方法一致，遥相呼应，稳定厚重，凸显出三弯腿劲挺灵秀。

明万历 《列女传》插图
鲍氏知不足斋刊本

Wanli Period of Ming Dynasty (1573-1620) Illustration of *Biography of Famous Women in Ancient China*, Zhibuzuzhai's edition from Family Bao

It's made of white dalbergia cochinchinensis, with a rectangular table top assembled with frames butted together by mitered corner bridle joints, and embedded flat central panel; edges are narrowed down widthways (also called ice-plate edge) and cut in a rabbet along bottom rims as borderlines; below the top panel installs a high girdled waist framed by extended parts of four legs at corners above the moulding; a short rod is installed in the middle of the waist to divide two embedded decorative panels carved with openwork of dragons holding up Chinese character "寿" (longevity) in a begonia-shaped window frame outlined with embossed borderlines on each side.

Below the girdled waist connects moulding, *Kun*-gate apron plates and four three-curved legs (cabriole legs) which bulge out a bit at upper parts, then tilt inwardly and turn into rolled-up leaf feet at end to perfectly exhibit the steadiness of the structure; curves of apron plates run smoothly along the bottom rims and connect to four legs at corners where cloud-shaped bas-relief is carved as copper corner protectors on bronze ware.

A bottom base is installed directly below four feet; the base top is constructed with assembled frames with cambered edges and double rabbets along the top and bottom rims, and a hard central panel; below the top panel connects a high girdled waist (higher than the one butted below the table top to reflect different layers of structure) with three short rods dividing two embedded decorative panels carved with openwork of floral patterns framed by begonia-shaped window and natural shaped borderlines; below the waist installs plates carved with *Kun*-gate shaped bas-relief.

This rectangular altar table is built in accurate structure, with table top and bottom base processed into the same style to maintain a steady and solemn appearance and to better set off the delicacy of three-curved legs (cabriole legs).

案

Narrow Table (An)

案是中国家具中历史甚为悠久的一类，品类亦多，在席地而坐的低型家具体系中已是主流，“拍案而起”，是书案；“举案齐眉”，是食案，案与人们日常生活息息相关。今人所谓的案，则是专指狭长形的一类家具，其腿足不位于四角，而是两端往里缩进，两头挑出，实际是从建筑梁架系统或壸门造型转化而来的，相应地形成了夹头榫结构和插肩榫结构。案中最经典者莫过于刀牙板平头案了，尺寸大小变化，大者可作画案，小者足充香案。更小的还有一种搁置案上的案上几，也是这种造型，简洁明快，不施雕琢，以拿捏到位的比例和优美的线条为装饰，虽然结体相同，但因细节变化的不同，呈现出千变万化的气象。本书收录的刀牙板平头案，就呈现出各不相同的美感，需要观者去细心品味。

Narrow tables constitute a time-honored category of classical Chinese furniture with a large variety of forms and designs, and have been widely pursued as early as the time when low-seat system dominated, which have been extensively reflected in idioms and set phrases, including “to pound the table and stand up” (referring to writing desk), “wife holding the table to the level of her brows while serving her husband” (referring to dining table); narrow tables have been greatly involved in people's daily lives; and today's so-called narrow tables specifically refer to a category of furniture of long and narrow appearance and with four legs indented below the top panel and two protruding short edges; in fact, T-bridle joint (clamp joint) and molded-frame bridle joint commonly seen on narrow tables are respectively originated from the truss structure of ancient Chinese architecture or the structure of *Kun*-gate; and the flat-edge narrow table with T-bridle joints (clamp joint) is the most classical kind; with varied sizes, it can be employed as drawing table (in big size), incense table (in small size), or even the table shelf (in smaller size); with such unadorned clean and simple structure, properly balanced scales and elegant silhouette, the narrow table can exhibit a kaleidoscope of variants; those flat-edge narrow tables with T-bridle joints (clamp joint) included in this book are perfect demonstration of its ever-changing beauty.

069

刀牙板小画案

Small Drawing Table with
T-bridle Joints (Clamp Joint)

88×57.5×79cm

画案由白酸枝制成，面镶瘿木。选料精良，色泽莹润，纹理流畅。

案长方，案面四面攒成，大边与抹头以格角榫连接，面心平装薄板，薄板下以穿带连接，穿带再与两大边连接。边抹做混面压边线线脚处理，弧面挺劲。

案面之下安四圆腿，四腿八挓，收分明显，增加下大上小的稳定感。圆腿与案面以高低榫连接。案面下、两腿间安刀板牙子，牙条与牙头一木连做，为一整料挖出。牙子边缘起阳线，以勾勒轮廓。

两腿之间侧面为梯子枨，梯子枨安装偏上，显得整体条案更有精神。梯子枨截面为鸭蛋圆，从不同角度看梯子枨的厚度也会不同，丰富了家具的细节。

此案造型简洁精炼，没有多余的雕饰与线脚处理，任其光素，更显气质非凡。

清　殷奇《春宫图》册
美国波士顿美术馆

Qing Dynasty (1644-1911)
Illustration of *Erotica Stories* (album)
By Yin Qi
Museum of Fine Arts, Boston

The drawing table is made of white dalbergia cochinchinensis of highly selected material and embedded with a burl central plaque with elegant colour and refined smooth veining.

This rectangular table top is constructed with assembled frames butted together by mitered corner bridle joints and embedded flat central panel, and is strained by sliding dovetail splines butted into long edges at the bottom; unadorned cambered edges are cut in a rabbet along the bottom rims as borderlines.

Below the table top installs four splayed legs with round sections by uneven double tenons to maintain the steadiness of the structure; unadorned one-piece straight apron plates are butted between legs by T-bridle joints (clamp joint) and outlined with embossed borderlines on front and back sides.

Between legs on two sides connects double ladder rails which are deliberately placed higher to raise the gravity centre so as to make the furniture appear more upstanding and dignified; ladder rails are carefully processed into cylindrical components of oval sections, therefore they shall appear in various widths from different angles of view, which subtly enriches decorative details.

This narrow table is of clean and simple structure, without any redundant carvings or mouldings, maintaining an unadorned appearance of extraordinary grace and beauty.

070

刀牙板平头案

Narrow Table with T-bridle Joints (Clamp Joint)

108×56×82cm

案长方，紫檀镶瘿木而成。案面四面攒边，面心平装优美的瘿木，瘿木色泽橙黄莹润，花纹细密旋转，瘿结蟠屈，似满面胡花。瘿木的淡雅与家具本身的深穆形成鲜明对比。边抹做冰盘沿压边线线脚处理。

案面下留出吊头安四腿，腿与案面高低榫连接。圆腿四腿八挓，特别向两侧收分明显，形成下大上小的稳定感。案面下两腿间安刀板牙子，刀板牙子修长素雅，不做雕琢，不做线脚处理。侧面两腿之间安梯子枨，梯子枨略高，更为精神。枨截面为鸭蛋圆，与腿的圆形截面相呼应。

此平头案体形小巧，比例协调，仅用简洁的刀板牙子、梯子枨和圆腿形成稳定坚固的结构，又形成简洁无饰的造型，结构与造型融为一体，为明式典型制式。

The narrow table made of red sandalwood, is constructed with assembled frames and an embedded flat burl plaque of bright orange colour, soft gloss, and swirls of refined veining, creating a mesmerizing illusion of tens of thousands of flowers in full bloom across the table top, forming a striking contrast between burl's elegance and red sandalwood's solemnity; cambered edges are narrowed down widthways (also called ice-plate edge) and cut in a rabbet along bottom rim as the borderline.

Below the top panel installs four splayed legs of round sections, indented and butted by uneven double tenons to maintain the steadiness of the structure; between legs connects straight and unadorned apron plates butted by T-bridle joints (clamp joints); Between legs on two sides connects double ladder rails which are deliberately placed higher to raise the gravity centre so as to make the furniture appear more upstanding and dignified; ladder rails are carefully processed into cylindrical components of oval sections to echo legs of round sections.

This flat narrow table of proper scales is built with unadorned apron plates of clean and simple beauty, ladder rails and legs to provide steady structure and also to exhibit a unified form of simplicity and elegance, hence a representative Ming-styled furniture embodying the perfect combination of structure and form.

明万历
杨之炯《蓝桥玉杵记》插图
建阳萧氏师俭堂刊本

Wanli Period of Ming Dynasty (1573-1620)
Illustration of *Story of A Jade Pestle at Bridge Lan*
By Yang Zhijiong
Shijiantang's edition from Family Xiao of Jianyang

071

刀牙板平头案

Narrow Table with T-bridle Joints (Clamp Joint)

133×68×83cm

平头案由黄花梨制成，黄花梨色泽淡雅温润，油性充足。

案长方，案面四面攒边，面心平装，边抹做冰盘沿压边线线脚处理。桌面下留出吊头装四腿，腿截面圆形，四腿八挓。案面下、两腿间安刀板牙子，牙子修长素雅，不起阳线，不做线脚变化。侧面两腿之间安梯子枨，枨截面为鸭蛋圆，饱满圆润。

刀牙板、梯子枨、圆腿平头案是典型的明式制式。此平头案比例修长，造型简洁，刀牙板、梯子枨既是连接案面和腿足的结构构件，又是塑造形态的造型元素，将结构与造型融合为一体。

It's made of yellow *Huali (Huanghuali)* wood of refined material with elegant colour and gloss, rich oiliness.

The rectangular table top is constructed with assembled frames and embedded flat central plaque; cambered edges are narrowed down widthways (also called ice-plate edge) and cut in a rabbet along bottom rims as borderlines; below the top panel installs four indented splayed legs of round sections to maintain the steadiness of the structure; between legs connects straight and unadorned apron plates butted by T-bridle joints (clamp joints) without any embossed borderlines; Between legs on two sides connects double ladder rails carefully processed into cylindrical components of oval sections.

Ming-styled narrow table is typically featured by apron plates butted by T-bridle joint (clamp joint), double ladder rails and cylindrical legs of round sections. This flat narrow table is of slender appearance, clean and simple structure; apron plates and ladder rails are structural components connecting table top and legs, and also essential elements sculpting the appearance of the furniture, hence a smooth combination of structure and form.

072

刀牙板画案

Drawing Table with T-bridle Joints (Clamp Joint)

196×86×83cm

画案黄花梨制，作平头案式，面长方，案面四面攒边，大边与抹头以格角榫连接，面心平装薄板，薄板之下安穿带，穿带再与两大边连接。边抹做冰盘沿压边线线脚处理。

案面之下留出吊头装四腿，腿与案面以高低榫连接。腿截面圆形，粗硕有力，四腿八挓，收分明显，以增加稳定性。案面下、两腿间安刀板牙子，牙条长直，牙头下坠，牙子裸素，不做任何修饰处理，显示出木材自然的纹理。侧面两腿之间安梯子枨，梯子枨偏上安装，增加平头案的气质。枨截面为鸭蛋圆，饱满圆润，从不同角度观察梯子枨，可得到不同直径的主观感受，丰富了家具的细节。

此件画案比例修长，造型简素雅致，不着修饰，又处处精炼，为明式家具之珍品。

The rectangular table top made of yellow *Huali* (*Huanghuali*) wood, is constructed with assembled frames butted together by mitered corner bridle joints and embedded flat central panel, and is strained by sliding dovetail splines butted into long edges at the bottom; cambered edges are narrowed down widthways (also called ice-plate edge) and cut in a rabbet along bottom rims as borderlines.

Below the top panel installs four splayed legs of round sections, indented and butted by uneven double tenons to maintain the steadiness of the structure; between legs connects straight and unadorned apron plates butted by T-bridle joints (clamp joints) without any embossed borderlines or decorations so as to clearly present natural veining of the material; Between legs on two sides connects double ladder rails which are deliberately placed higher to raise the gravity centre so as to make the furniture appear more upstanding and dignified; ladder rails are carefully processed into cylindrical components of oval sections, therefore they shall appear in various widths from different angles of view, which subtly enriches decorative details.

This flat narrow table is of slender appearance, clean and elegant structure without any decorations, perfectly demonstrating the essence of classical Ming-styled furniture.

073

卷云纹平头案

Narrow Table with Apron Plates Carved in Cloud-shaped Design

186×56×85cm

平头案由花梨木制成，花梨木色泽淡雅莹润，纹理细腻优雅。

平头案长方，案面四面攒边，大边与抹头以格角榫连接。面心平装薄板，薄板之下安穿带，穿带再与两大边榫接。边抹做混面压边线线脚处理，不同于常规之处在于，边线为皮条线加洼儿，做法独特。

案面下留吊头，安四腿，腿与案面以高低榫连接。腿截面为方形混面，正面起宽皮条线加洼儿，与案面边抹处理方法呼应。四腿八挓，收分明显，以增加下大上小的稳定感。案面下、两腿间安花牙子，牙条素直，两端起委角曲线，翻转到牙头，成卷云纹。花牙子边缘起阳线，以勾勒轮廓。

此平头案比例修长，做工精湛，桌面边抹的弧面压皮条线加洼儿的做法较为独特， 腿足前面起的皮条线加洼儿，与桌面边抹处理方法相一致，使得整件家具协调一致，气质独具。

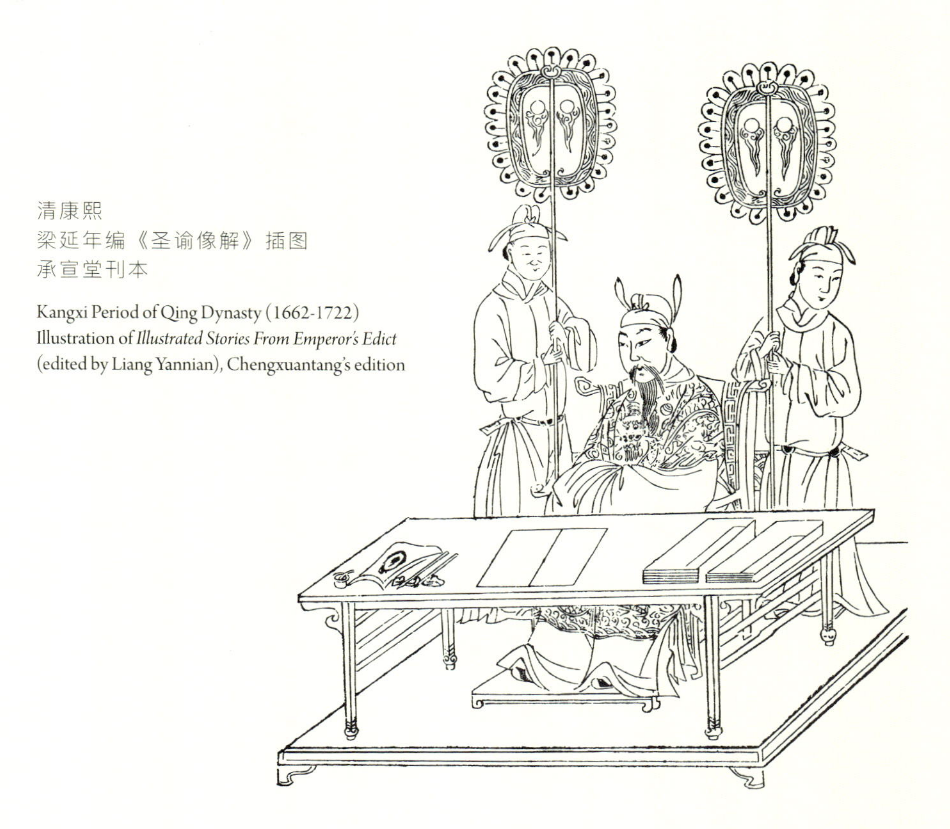

清康熙
梁延年编《圣谕像解》插图
承宣堂刊本

Kangxi Period of Qing Dynasty (1662-1722)
Illustration of *Illustrated Stories From Emperor's Edict* (edited by Liang Yannian), Chengxuantang's edition

This flat narrow table is made of *Huali* wood of refined material with graceful colours and smooth veining.

The rectangular table top is constructed with assembled frames butted together by mitered corner bridle joints and embedded flat central panel, and is strained by sliding dovetail splines butted into long edges at the bottom; unadorned cambered edges are uniquely cut in an embossed belt-lines (wider than the usual borderlines) gouged out a round low-lying groove in the middle.

Below the top panel installs four splayed legs of chamfered square sections, indented and butted by uneven double tenons to maintain the steadiness of the structure; An embossed belt-line (wider than the usual borderline) gouged out a round low-lying groove in the middle is carved on the exterior of each four legs to echo edges of the table top; between legs connects straight and unadorned apron plates outlined by embossed borderlines and corner plates carved in abstract cloud-shaped pattern.

This flat narrow table is of slender appearance, and the embodiment of highly skilled craftsmanship of ancient Chinese; its unique embossed belt-lines (wider than the usual borderlines) gouged out a round low-lying groove in the middle on exteriors of four legs echo the table top's edges, and maintain a distinct unified style.

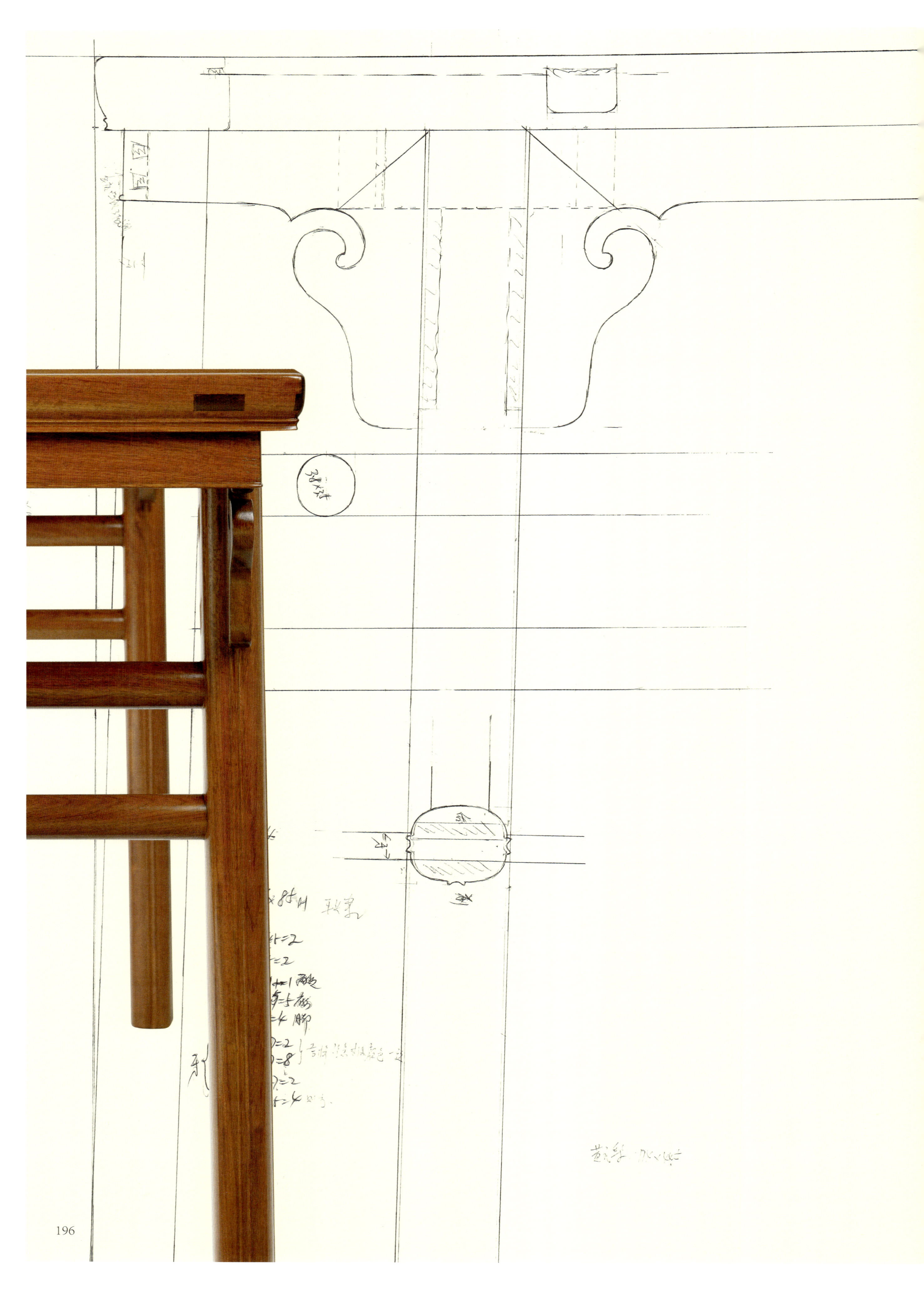

清　郎世宁《乾隆博古图》　故宫博物院藏

Qing Dynasty, *Emperor Qianlong Appreciating Antiques*
By Giuseppe Castiglione (Lang Shining),
Collection of the Palace Museum

074

刀牙板瓜棱腿翘头案

Narrow Table with Raised Short Edges, T-bridle Joints (Clamp Joint) and Legs of Melon-shaped Sections

118×42×81.5cm

翘头案由花梨木制成，选料讲究，色泽淡雅，纹理流畅。

案长方，案面为独板，两短边拍抹头，抹头起翘，小巧精雅，精神抖擞。独板立面做混面压边线线脚处理。

案面下留吊头，装四腿，腿与案面以高低榫连接。四腿八挓，收分明显，以增加稳定感。腿截面为瓜棱形，线脚变化丰富。案面下两腿间安刀板牙子，一木连做，牙条纤细，两侧翻转出曲线与牙头连接。刀板牙子边缘起阳线，以勾勒轮廓。

侧面两腿之间安梯子枨，梯子枨偏上安装，更显气质。梯子枨截面亦为瓜棱形，与案腿相呼应。

此翘头案体形小巧，比例合宜，细节精致，气质独特。

This narrow table with raised short edges is made of *Huali* wood of refined material with graceful colours and smooth veining.

The rectangular one-piece top panel is butted with two delicate raised short edges; and two long cambered edges are cut in a rabbet along bottom rims as borderlines.

Below the top panel installs four splayed legs of melon-shaped sections, indented and butted by uneven double tenons to maintain the steadiness of the structure; between legs connects straight and slender one-piece apron plates by T-bridle joints (clamp joints); unadorned plates have smooth curves at corners and are outlined with embossed borderlines.

Between legs on two sides connects double ladder rails which are deliberately placed higher to raise the gravity centre so as to make the furniture appear more upstanding and dignified; ladder rails are of melon-shaped sections to echo the style of four legs.

This narrow table is built in small size and proper scale, with delicate details and unique character.

075

对卷云纹带托泥翘头案

Narrow Table with Raised Short Edges and Apron Plates Carved in Cloud-shaped Design

206.6×63.2×83cm

翘头案以黄花梨制成，阔而长，可充画案之用。

面框冰盘沿立墙，微打洼，与腿足呼应。两头为与抹头一木连做的翘头，造型小巧而直翘，家具行所谓"拇指翘"，大案小翘，相得益彰，显示了匠作非一般的把控能力。活面做法，案面边抹与牙板以栽榫相合。

牙板甚厚，既便设榫，又可保证拆卸后的牢固度。前后牙板间设有托带辅助。牙头为对卷云纹式，不设一线。腿足看面微打洼，转角处起委角线。侧向两腿足间装壸门券口牙板，线条流畅，造型古朴，有趣的是两竖牙条并不落在托泥上，而是另设弧弯的横枨，悬空而设，交于腿足，形若两个相抵的卷珠纹，这种形式的花枨为苏北家具突出特征。

此案尺寸硕大，细节处处特殊，整体和谐统一，榫卯得宜，制作精细。侧面券口牙板竖牙条上翻花叶的造型，与明晚期尤其是万历时期柜格券口上所见相近。

This narrow table with raised short edges is made of yellow *Huali (Huanghuali)* wood of refined material; it can be used as drawing table due to its longer and wider top panel.

Long edges of the top panel are narrowed down widthways (also called ice-plate edge) and gouged out a round low-lying groove lengthways to echo four legs; two delicate raised short edges are also called "thumb-raised" edges by connoisseurs, creating a contrast between the long and wide top panel and small raised edges, reflecting craftsman or owner's extraordinary control on scale and structure; apron plates are butted into edges of table top by hidden splines.

Apron plates are thick enough to bear hidden splines and also maintain the steadiness of the structure; front and back apron plates are connected on two sides; corner joints of apron plates are carved in opposite cloud-shaped patterns without any borderlines; exteriors of four legs are slightly gouged out a round low-lying groove lengthways to echo the table top's edges; corners of four legs are chamfered and cut in two rabbets lengthways; between legs on two sides connects *Kun*-gate arch plates of beautiful curves and archaic style below the table top, yet two vertical plates are directly butted with an extra curved rail carved in the shape of a pair of rolled-up pearls instead of the foot rail on each side, which is the distinct feature of classical furniture produced in Northern Jiangsu Province.

This narrow table is built in large size with extraordinary details and properly employed mortise-and-tenon joints, maintaining a unified style. The floral silhouette of vertical arch plates on two sides below the top panel bears a close resemblance with those arch plates installed in cabinets of Wanli Period in the late Ming Dynasty.

清早期　佚名《清宫珍宝皕美图》插图

Early Qing Dynasty (1644-1911)
Illustration of *Two Hundred Treasures from Royal Collection of Qing Dynasty* (album)
By Anonymity

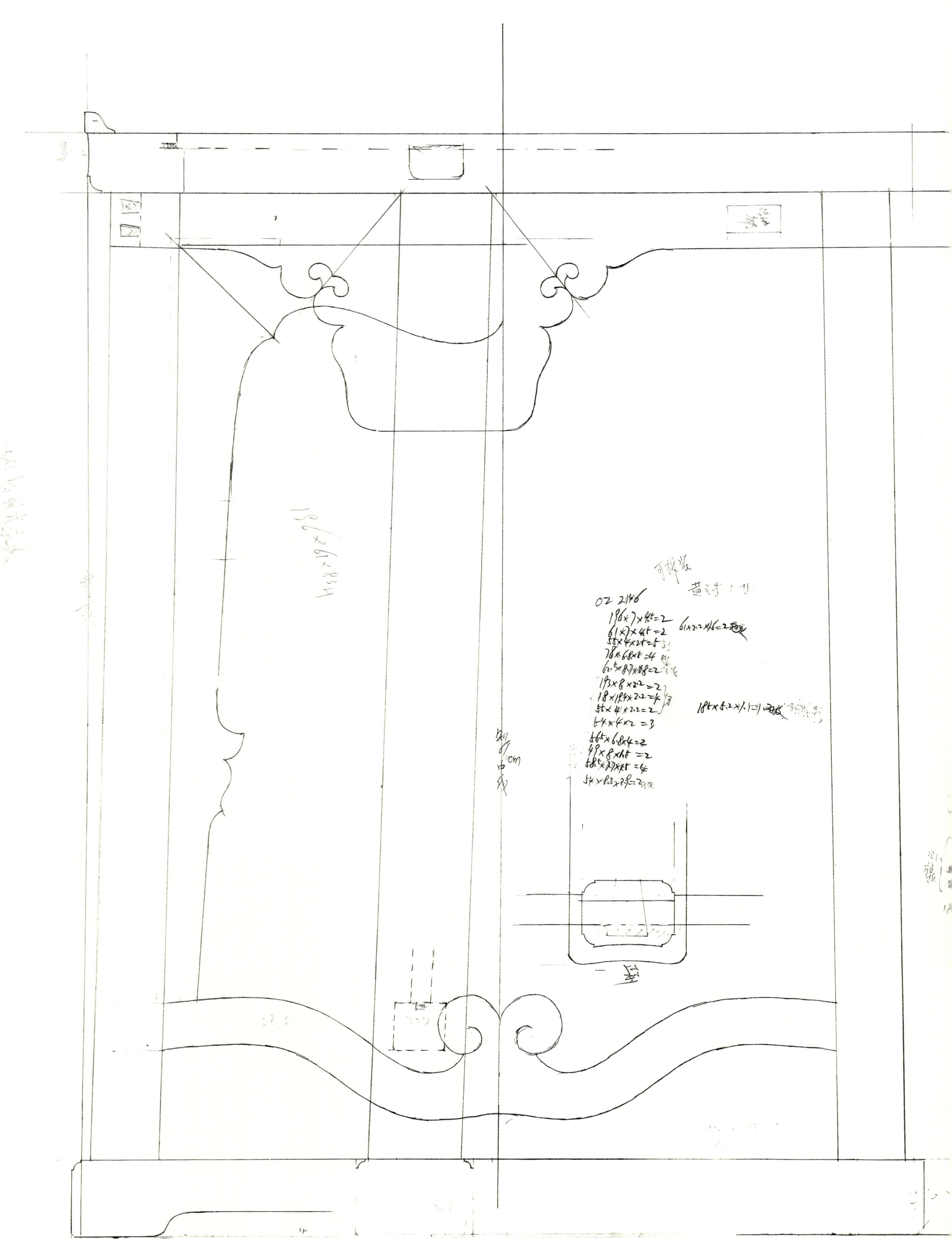

076

螭龙纹带托泥翘头案

Narrow Table with Raised Short Edges, Foot Rails and Baffle Panels Carved with Openwork of Dragons

215×48×86cm

此翘头案由花梨木制成，花梨木色泽淡雅莹润，纹理细腻流畅。

翘头案长方，案面四面攒边，面心平装薄板，边抹做冰盘沿压边线线脚处理。抹头起翘头，翘头向外延伸，曲线优雅。

案面之下留出吊头，装四腿，四腿八挓，收分明显，增加稳定感。腿截面线脚丰富，内侧平直，外侧两端先起阳线，再做混面，中间打洼。案面下、两腿间安云头牙子，牙条长直无饰，两端向下翻转形成云头纹饰。牙子边缘起阳线，以勾勒轮廓。

侧面两腿以横枨和托泥围合成板腿，中间装雕螭龙纹挡板。

此翘头案体量较大，用料粗硕大气，线脚处理细腻自如，加之适当的雕刻装饰，简洁中透着细致秀雅。

This narrow table with raised short edges is made of *Huali* wood of refined material with graceful colours and smooth veining.

The rectangular table top is constructed with assembled frames and embedded flat central panel, with edges narrowed down widthways (also called ice-plate edge) and cut in a rabbet along bottom rims as borderlines; top panel is butted with two raised short edges of elegant and protruding curves.

Below the top panel installs four indented splayed legs of special-shaped sections to maintain the steadiness of the structure; exteriors of four legs are outlined with embossed borderlines, with each cambered surface gouged out a round low-lying groove lengthways in the middle; between legs connects long and straight apron plates outlined with embossed borderlines with corner joints carved in cloud-shaped pattern.

A baffle panel framed by a horizontal rail and a foot rail between two legs is installed below the table top on each side, which is carved in exquisite openwork of dragon pattern.

This narrow table is built in larger size, with solid components and delicate decorations, especially these extraordinary carvings, exhibiting a beauty of simplicity and gracefulness.

077

灵芝纹翘头案

Narrow Table with Raised Short Edges, Apron Plates Carved in Ganoderma-shaped Design and Baffle Panels Carved with Openwork of Ganoderma

275×60×88/93.5cm

翘头案由白酸枝制成，选料讲究，色泽淡雅莹润，纹理流畅细腻。

案长方，案面四面攒边，面心平装薄板，边抹做混面压边线线脚处理。抹头起翘头，翘头曲线圆转自如，优雅秀丽。

案面下留吊头，安四腿。四腿八挓，收分明显，形成稳定感。腿足下端向外微撇，故称“香炉腿”。腿截面做混面压边线线脚处理，混面圆浑。

案面下、两腿间安云头灵芝牙子，牙条长直光素。两端向下翻转出云头牙子，云头之上再阴刻灵芝纹。牙子边缘起阳线，以勾勒轮廓。

侧面两腿间以横枨和托泥围合出板腿，中间装灵芝草纹挡板。

此翘头案修长粗硕，用料厚实，塑造出大案重器的大气感。

明万历
《御世仁风》插图
凤阳刊本

Wanli Period of Ming Dynasty (1573-1620)
Illustration of *Collected Stories of Emperors*, Fengyang edition

清早期　佚名《清宫珍宝皕美图》册

Early Qing Dynasty (1644-1911)
Illustration of *Two Hundred Treasures from Royal Collection of Qing Dynasty* (album)
By Anonymity

The narrow table is made of white dalbergia cochinchinensis of refined material with elegant colour and smooth veining.

The rectangular table top is constructed with assembled frames and embedded flat central panel, with cambered edges cut in a rabbet along bottom rims as borderlines; top panel is butted with two raised short edges of elegant and protruding curves.

Below the top panel installs four indented splayed legs to maintain the steadiness of the structure; four legs curve slightly at end, as "censer legs"; exterior of each leg is polished in cambered surface and cut in two rabbets lengthways as borderlines.

Between legs connects apron plates carved in shapes of cloud and ganoderma; plates outlined with embossed borderlines are slim and straight with corner joints carved in the shape of abstract cloud pattern and with intaglio of ganoderma motif.

A baffle panel framed by a horizontal rail and a foot rail between two legs is installed below the table top on each side, which is carved in exquisite openwork of ganoderma pattern.

This narrow table is of long and sturdy appearance, made of large refined material, to perfectly exhibit a dignified solemnity.

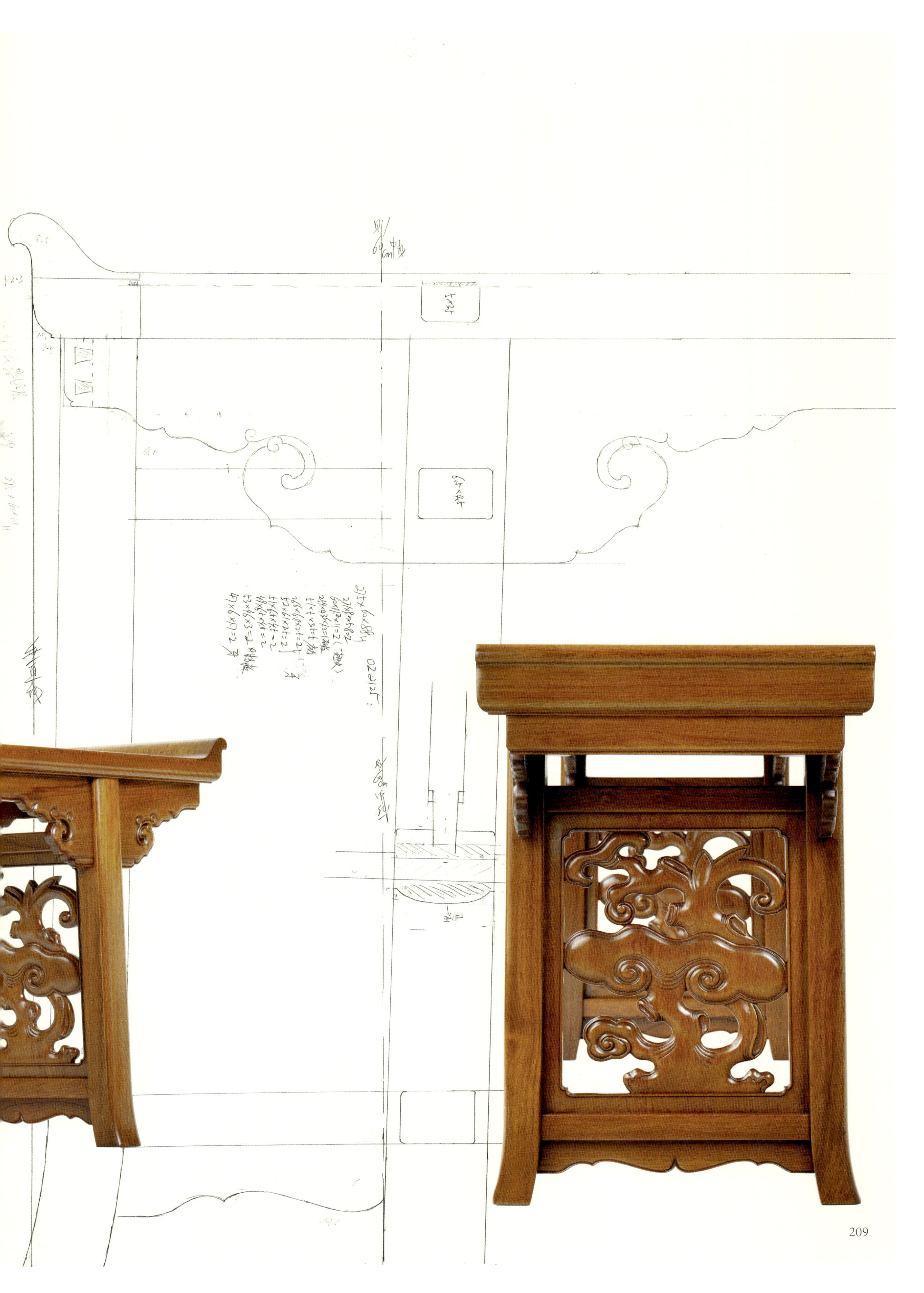

078

卷云纹插肩榫箭腿翘头案

Narrow Table with Raised Short Edges, Molded-frame Bridle Joints, Apron Plates Carved in Cloud-shaped Design and Sword-shaped Legs

108×33×81cm

翘头案由黄花梨制成，黄花梨色泽淡雅，纹理流畅细腻，油性充足。

案长方，案面独板制成，即所谓一块玉。抹头与翘头一木连做。抹头起翘，翘头小巧精神。边抹做冰盘沿压边线线脚处理。

案面下留出吊头，安四腿，腿为箭腿，正面向两侧收分明显，以增加稳定性。腿与案面以高低榫连接。箭腿自上竖直而下，逐渐收细，于末端外翻出马蹄形，箭腿边缘起阳线，至马蹄位置向内翻出云头纹。马蹄之下再叠两层台座，形成完整的腿足。

案面下、两腿间安壶门云头牙子，一木连做，牙条做壶门曲线，曲线自花心向两侧翻转，在两侧向下翻转出云头，与腿足以插肩榫连接。牙子边缘起阳线，随着壶门曲线、云头曲线翻转，与腿边缘的阳线相接。

侧面两腿之间安梯子枨，枨截面为平面委角线脚。

此翘头案小巧精致，精神抖擞，细节处理精益求精，经得起推敲。

The narrow table is made of *Huali (Huanghuali)* wood with elegant colour, gorgeous veining, and rich oiliness.

The rectangular one-piece top panel is assembled by two delicate frames with raised short edges; edges are narrowed down widthways (also called ice-plate edge) and cut in a rabbet along bottom rims as borderlines.

Below the top panel installs four splayed sword-shaped legs, indented and butted by uneven double tenons to maintain the steadiness of the structure; four legs outlined with embossed borderlines run straightly downward, taper gradually and turn to hoof feet carved with delicate cloud-shaped pattern at end; four small double-decked foot pads are butted into hoof feet.

Between legs connects one-piece *Kun*-gate apron plates outlined by embossed borderlines with smooth curves undulating from the centre towards two corner joints and turning into cloud-shaped carvings; four legs are butted with apron plates by molded-frame bridle joints; between legs on two sides connects double ladder rails of rectangular (with dented corners) sections.

This narrow table is built in small size, with exquisite decorations and details that are worth stance of looking.

079

卷云纹插肩榫箭腿翘头案

Narrow Table with Raised Short Edges, Molded-frame Bridle Joints, Apron Plates Carved in Cloud-shaped Design and Sword-shaped Legs

147×39×83cm

翘头案以黄花梨制成，面镶金丝楠老料。

案面攒框，冰盘沿线脚挺拔硬朗，下方压阳线，小巧可人。两端设小翘头，短促有力，即行家所谓的"一指翘"，形成了丰富的细节变化，也起到阻挡物品两头滑落的实用功能。案面心为一整块金丝楠老料，色泽如陈年老绢，纹路如一幅优美的宋画山水，有风起云涌、巨浪滔天之势，如菡萏之发，如春山之烂漫，气象万千。壶门牙板，曲线挺拔而优美，卷云纹牙头，小巧可爱，牙板和腿足皆沿边起阳线强调。

插肩榫箭腿式，腿足中间起一炷香细线，两端压边线。腿足中端外翻花牙，下端向两端外撇如马蹄，浮雕如意云头，云头尖处上沿为一炷香线。下承一木制成的梯形垫木，形如筌蹄、莲台。腿间侧面连以委角方形梯子枨。

插肩榫结构的翘头案，相对夹头榫少见，但是往往别发机杼，有意想不到的造型之美，譬如此例，就是很成功的设计。

It's made of yellow *Huali (Huanghuali)* wood of refined material, and embedded with an old golden nanmu (Phoebe sheareri) central panel; two exquisite raised short edges also called "one-finger-raised" edges by connoisseurs are butted into the top panel on two sides to create more variations in detail and also prevent objects slipping off the table; the embedded central plaque is made of an one-piece old golden nanmu (Phoebe sheareri) with solemn colour and golden sheen like aged silk, refined natural veining swirling and flowing like a gorgeous landscape painting of the Song Dynasty (960-1279) or the ever-changing cloud in the sky, surging tides in the ocean, blooming flowers in springtime; below the table top installs *Kun*-gate apron plates outlined by embossed borderlines with graceful and strengthful curves and corner joints carved in delicate cloud-shaped patterns.

Four sword-shaped legs are butted into apron plates by molded-frame bridle joints and outlined with embossed borderlines, with an extra embossed line (incense line) carved lengthways in the middle of each leg; sword-shaped silhouettes turn to floral pattern in the middle and tilt outward as hoof feet at end carved with delicate bas-relief of cloud-shaped pattern and butted with four trapezoid foot pads carved in the style of lotus Sumeru seat; between legs on two sides connects double ladder rails of rectangular (with dented corners) sections.

Narrow tables assembled by molded-frame bridle joints are rarer than those by T-bridle joints (clamp joints), however, they usually demonstrate a unique beauty beyond expectation.

炕几

Tray Table

矮型承具在席地而坐时代，是主要的家具门类，在垂足而坐盛行的高型家具体系中，处于次要地位，但存量亦不少，其原因众多，诸如：北方卧具以炕为主，则配合炕的炕桌大行其道；日常生活中，炕桌等家具小巧方便，收纳携带便利，在比较随意和临时的场所，无疑是更好的选择。传世的矮型承具，广义多名以炕桌，若要细分，将宽近长三分之二或更多者名炕桌（狭义），将狭长桌型结构者名炕几，案型结构者名炕案，以示区别。本书收录炕几二件，皆是板足造型，是此类中造型简练者，搁置罗汉床、榻上，或充客厅、书房矮几之用。板足几又被俗称为"炕琴""琴几"，不一定非要弹琴之用，但却似乎以琴相伴，方显这类家具的清雅。另录一件炕桌，面为棋盘，是可供对弈的棋桌。

In the era when people sat on mats laid on the ground, low-top-panel furniture-to-present/carry dominated China; however, during the period when high-seat furniture prevailed, they fell to the subordinative position with still a considerable number as a result of various reasons, including: the heatable brick bed (*Kang*) widely used in North China mostly requiring the company of tray tables; in daily lives, small and portable tray tables becoming a better choice for casual and temporary occasions. Most exsisting low-top-panel furniture-to-present/carry can be refered as tray table generally; in narrow sense, those with width-length ratio of 2:3 or more are referred as tray table, those with narrower width are referred as end table (*Ji*), those built in the structure of narrow table are referred as narrow table (*An*). This book included two tray tables with plank feet, representing the simple and clean structure of tray table; they can be placed on arhat bed and couch, or served as short end table (*Ji*) in living room or study. Plank-feet tray tables are also called "*Guqin* (an ancient Chinese seven-stringed plucked instrument) table"; although they don't necessarily go with musical instrument, such combination truly better reveals the elegance of the furniture. Another chess/go table is also included in this book.

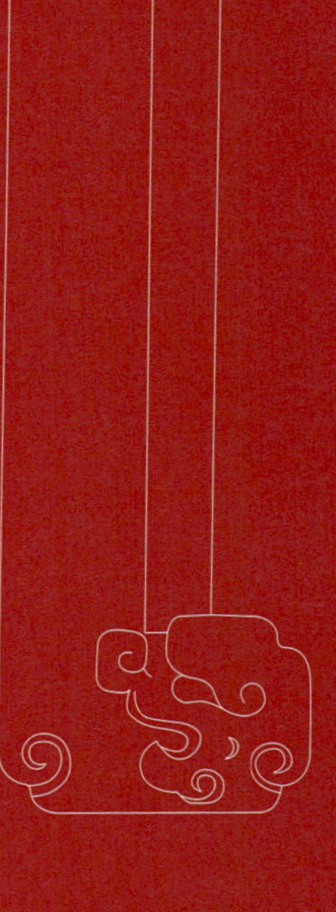

080

卷几式炕几

Tray Table with Plank-feet

152×34×37cm

炕几由花梨木制成，花梨木色泽淡雅温润，纹理细腻流畅。

炕几由三块厚板直角相交而成，几面长直，立面做素混面线脚处理。几面两端向下翻转，与侧面的板腿圆角相接。几面下紧贴一条垛边，垛边边缘起阳线，两侧亦做曲线，与板腿的竖向垛边连接。垛边的用意在于增加面板外观的厚度，增加细节处理变化。

两侧板腿略往外撇，与几面连接大于直角，出挓，形成上小下大的稳定感。板腿落地之处向内翻转，类似卷书的做法，立面在卷书造型基础上浅浮雕回首螭首纹。板腿外缘落地之处，内翻云头，以呼应螭首纹。沿着侧板腿内缘亦安垛边，上与几面的垛边圆角相接。

此炕几造型优雅，用材粗硕，圆浑无棱角，气质静穆典雅。

The tray table is made of *Huali* wood with elegant colour and smooth veining.

This tray table is assembled by three major planks with unadorned cambered edges; table top bends down on two sides and connects to two plank-feet at rounded corners; three extra lining panels are also butted together at rounded corners and directly attached to interiors of three major planks with slightly intended edges outlined by embossed borderlines so as to increase the thickness of the planks visually and to add more variant details.

Two plank-feet slightly lean outward to maintain the steadiness of the structure, and roll up inwardly at end resembling scroll-feet in style; front facades of plank-feet tips are carved with decorative bas-relief of dragon head design; the exteriors of foot tips on two sides are carved with cloud-shaped pattern to coincide with dragon head design; extra lining panels are also attached to the interior of two plank-feet and butted to the one below the table top.

This tray table of elegant structure is made of large and solid material and processed to have rounded corner joints, exhibiting a graceful solemnity.

081

板足炕几

Tray Table with Plank-feet

146×35×38cm

炕几由花梨木制成，精心选料，色泽淡雅润泽，纹理细腻流畅。

炕几由三块厚板直角相交而成，几面长直，立面做素混面线脚处理，混面圆浑饱满。几面两端向下出曲线，与侧面板腿圆角相接。

侧面板腿，上半部分挖去板内侧，下半部分挖去板外侧，至腿下部稍稍向外翻转，略向上翘起。板腿上半部分中间挖出一孔，略如覆瓦，造型古朴。

此炕几造型简洁洗练，仅有三块厚板榫接而成，用材重硕，线脚圆浑饱满，气质沉稳娟秀，独具一格。

见王世襄《明式家具研究》乙15，页81。

The tray table is made of *Huali* wood of highly selected material with elegant colour and smooth veining.

This tray table is assembled by three major planks with unadorned cambered edges at right angels; table top bends down on two sides and connects to two plank-feet at rounded corners.

Two plank-feet are gouged away the inner upper parts and the outer bottom parts to leave foot tips slightly rolled up, creating a unique silhouette variation; an archaic tile-shaped hole is bored in the middle area on each side.

This tray table is built in clean and simple structure, assembled by three planks made of large and solid material, exhibiting a unique beauty of elegance and tranquility.

Reference to *Connoisseurship of Chinese Furniture: Ming and Early Qing Dynasties* by Wang Shixiang, B15, page 81.

082

矮方几式棋桌

Square End Table-Styled Chess/Go Table

56×56×13cm

棋盘由紫檀制成，紫檀色泽沉穆，牛毛纹满密，油性充足。

棋盘方正，案面四面攒边，边抹外边缘略高起，做拦水线。边抹做冰盘沿压边线线脚处理。面心平装，面心两块，再做四面攒边，心板上阴刻围棋、象棋棋盘，面心可更换，以满足不同棋类需要。

案面之下收束腰，束腰上开鱼门洞式开光三组，开光边缘起阳线以勾勒轮廓。束腰之下接四腿和牙子，四腿和牙子略向外膨出。四腿竖直，末端略向内翻，与管脚枨格角相接。管脚枨之下安小足。束腰下两腿间安牙条，牙条中间部分略下垂，向上翻出卷草纹，两侧与腿圆角相接。腿足下端安管脚枨，管脚枨做成罗锅式，罗锅拱起部分小巧挺劲，与牙条下垂部分相呼应。

此棋盘体量精巧洗练，比例协调，适当的线脚和曲线处理，增加了丰富的细节。

The chess/go table is made of red sandalwood with dark and solemn colour, visible ox-hair and pinhole veining (traces of capillary tubes of the plant), and rich oiliness.

The square chessboard is constructed with two embedded flat central plaques and assembled frames outlined by embossed retaining lines with unadorned cambered edges narrowed down widthways (also called ice-plate edge) and cut in with a rabbet along the bottom rims; two embedded central plaques constructed with assembled frames, are respectively carved with chess board for the game of go and Chinese chess in intaglio on each, which can be replaced as needed.

Below the top panel connects a girdled waist carved with slim triple-fish traceries outlined with embossed borderlines in the middle areas on four sides; below the girdled waist installs four legs slightly bulging out for a start and then running a bit inward at end to connect with foot rails by mitered corner bridle joints; four small foot pads are butted below four corners of foot rails; between legs connects apron plates at rounded corners, with middle parts hanging down a bit and carved with floral patterns; foot rails are processed in exquisite hunched style with straight and strong curves to echo the middle hanging parts of apron plates.

This chess/go table is built in simple delicacy, with proper scale , clean and neat mouldings and curves to add more exquisite details.

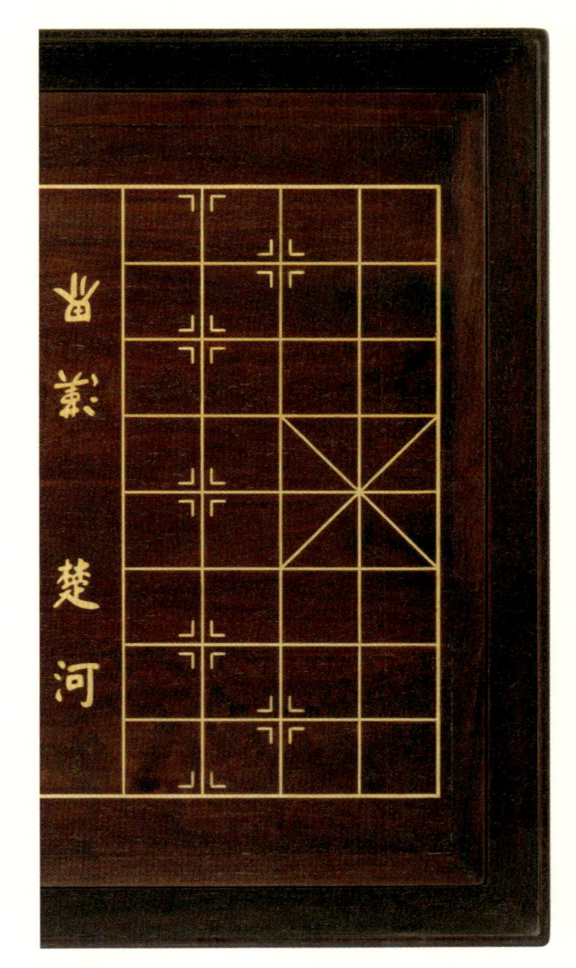

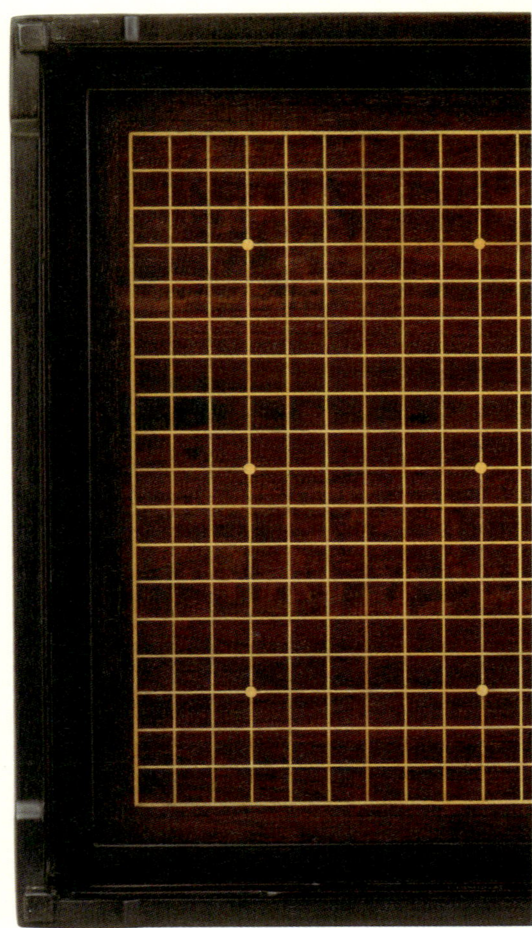

OU

FURNITURE COLLECTION OF FAMILY OU

VOLUME II

Compiled by Ou Shengchun
Published by Zhongshua Book Company

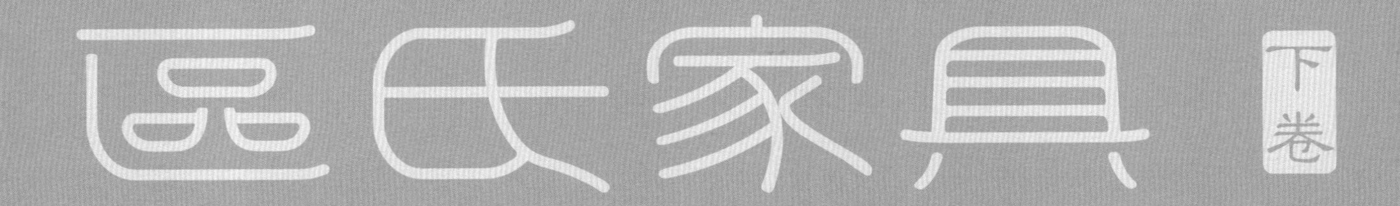

区胜春 著

中华书局

卧具

Furniture-to-rest-in

李笠翁《闲情偶寄》"器玩部·制度第一·床帐"云："人生百年，所历之时，日居其半，夜居其半。日间所处之地，或堂或庑，或舟或车，总无一定之在，而夜间所处，则止有一床。是床也者，乃我半生相共之物，较之结发糟糠，犹分先后者也，人之待物其最厚者，当莫过此。"卧具是各类家具符合人类最根本需求的，可以没有椅凳柜架，但不可不设卧具。也正是因此，卧具是各类家具中最早发展和成熟的，湖北荆门包山楚墓出土的黑漆大床，带围栏，可折叠，非常机巧，就是最好的印证。明清时期常见的卧具主要有架子床、罗汉床、榻等，其中后两者是兼有坐具功能的。随着今天起居方式的改变，古代卧具在现代居室的应用多被人忽略。其实这是一种情趣别致，具有深刻文化内涵且使用便利的器具，只有时常接触者，才知其佳味。

According to *Casual Literary Notes: Utensils—Bed & Drapery* written by Li Liweng (famous litterateur, dramatist and aesthetician in the late Ming Dynasty, 1611-1680), "throughout one's life no more than one hundred years, people spend half of the time in the day and half of it at night; in daylight, they dwell in the halls or rooms, on the boat or in the carridge as they wish; yet at night, people have no other places to go except for the bed; therefore, I shall spend half of my life with my bed, longer than anyone else, even comparing with my wife; naturally, I do reserve deepest affection for the bed." As the most indispensable classical Chinese furniture, other than chairs, stools, cabinets and shelves, furniture-to-rest-in is built to satisfy people's most primary needs, therefore, among all the furniture, it appeared and fully developed in the earliest time in ancient history like the black-lacquered baffled folding bed unearthed from the mausoleum of State Chu of Warring States Period (475 B.C.-221 B.C.) at Baoshan Hill, Jingmen City, in Hubei Province. Canopy bed, arhat bed, and short-legged couch are commonly seen in the Ming and Qing Dynasties and the latter two kinds can also serve as seating furniture. The introduction of ancient furniture-to-rest-in to modern interior settings are largely overlooked; actually it could turn out as an elegant interest of profound cultural connotation and functional convenience to those who have frequent contact and long-time companionship with them.

床

Bed

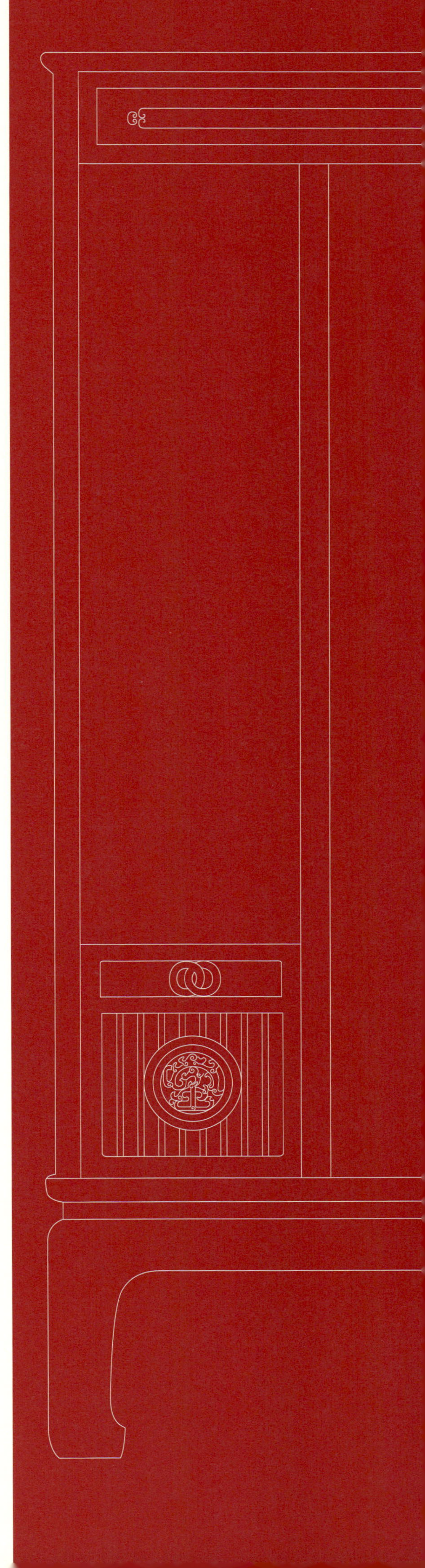

《释名》曰："人所坐卧曰床。床，装也，所以自装载也。"河南信阳长台关楚墓出土黑漆彩绘大床，是较早的卧床实例。明清时期，这种四柱或六柱架子床成为最主流的卧具，南床北炕，渐成习惯。架子床四面透风，又可挂设蚊帐、纱帐，上有仰尘，形成一个包围的心理空间，起居其中，安全感顿生。而周围的围子，透雕花纹，光线下各种投影。倘如《遵生八笺》所记，辅以花纸帐，挂壁瓶，插花卉，满床生香，却是一番别有情趣的旖旎光景了。罗汉床本是坐具，又可供小憩，最受文人青睐，搁置书房，不失雅致，或摊书卷、古董其上，随意坐卧赏鉴。三面围子，或为素板，欣赏自然而成的树木纹路，若山岚流水，或雕琢镶嵌，各尽能事，繁简不一，那就看主人的喜好了。

According to *Definition of Things and Matters* (written by Liu Xi in the late Eastern Han Dynasty, 25-220), "where people sit or rest in is called bed; bed, also meaning to load, indicates its function as loading." The black-lacquered canopy bed painted with colored drawings unearthed from the mausoleum of State Chu of Warring States Period (475 B.C.-221 B.C.) at Changtai Pass, Xinyang City, in Henan Province is an exisiting early example of ancient bed. By the Ming and Qing Dynasties, such four-post or six-post canopy bed became the major type and constituted the general concept of "wooden bed used in the south and heatable brick bed used in the north". The canopy bed is exposed between posts where it can hang curtain or crepe drapery, covered by a canopy to create an enclosed space substantially and psychologically so as to give a sense of security, and fenced up with baffle panels carved with openworks of various motifs allowing for light to come through and drop ever-changing shadows. According to *Eight Notes for Health Preserving* (written by Gao Lian in the Ming Dynasty 1368-1644), the canopy bed then were decorated with stained paper drapery and hanging wall-vases full of fresh flowers giving off sweet fragrance permeating the bed. The arhat bed is originally considered as seating furniture, and can be used while people take a nap; it's quite popular among literati as the arhat bed can be placed in the study to create elegant enviroments and also serve as lounge couch where people can appreciate antique objects or read books. The arhat bed is built with baffle panels on three sides either carved and inlaid with decorations or left unadorned to display the material's natural texture and veining as the owner likes.

083

独板围子罗汉床（附炕桌）

Arhat Bed with Three One-piece Panels
(with a Tray Table)

208×108×48/76cm

罗汉床由黄花梨制成，精心选料，纹理优美，色泽温润，油性充足。

罗汉床三面围子，皆独板黄花梨制成，且精选纹理优美的木料，有风起云涌之势，麦穗纹满布。围子边缘做素混面，边角做圆角，凸显罗汉床内敛儒雅的气质。床面四面攒边，面心编席面软屉。边抹做混面压边线，混面饱满简单，与围子相呼应。

床面下收矮束腰，束腰下接粗硕四腿，腿线条挺直，足端内翻马蹄。两腿之间装窄长的素直牙条，不做任何装饰，与整件罗汉床风格一致。牙条和四腿边缘起细阳线，修饰轮廓。值得关注的是，束腰与牙条一木连做，且矮束腰和窄牙条相搭形成的厚度，才与粗硕的四腿相配，此处可见设计之独到。

此件罗汉床秀雅修身，以长直线条为主要设计元素，贯穿于三面围子、床面、束腰、牙条和四腿，使得整件家具一致统一，一气呵成。

与罗汉床成套的束腰马蹄腿炕桌，沿用了罗汉床的长直线条为设计元素，造型亦洗练简洁，与罗汉床相呼应。

It's made of yellow *Huali (Huanghuali)* wood of highly selected material with gorgeous veining, elegant colour and rich oiliness.

The bed board is fenced up with one-piece baffle panels on three sides made of highly selected yellow Huali (*Huanghuali*) wood with graceful natural veining resembling to scudding clouds blown by wild wind and dense wheat veining scattered in between; unadorned cambered edges and rounded corners reflect arhat bed's subtle elegance; the bed board is constructed with assembled frames, an embedded soft woven central cushion and simple plain cambered edges with a rabbet along the bottom rim as borderline to echo the minimalist style of baffle panels.

Below the bed board installs a short girdled waist and four straight and stout legs with inwardly rolled up hoof feet at end; unadorned narrow apron plates are butted between legs to maintain the style of simplicity; apron plates and legs are outlined with embossed borderlines; and the girdled waist and apron plates are processed out of same material; uniquely, the short girdled waist and the narrow apron plates together create a visual horizontal weight so as to balance the four short stout legs.

This arhat bed is of a slender and elegant appearance depicted basically by long and straight linear structure on three baffle panels, bed board, girdled waist, apron plates and four legs to maintain a unified style.

The matched tray table which comes along with the arhat bed is built with the same girdled waist and hoof feet of linear structure to echo the minimalist style of the bed.

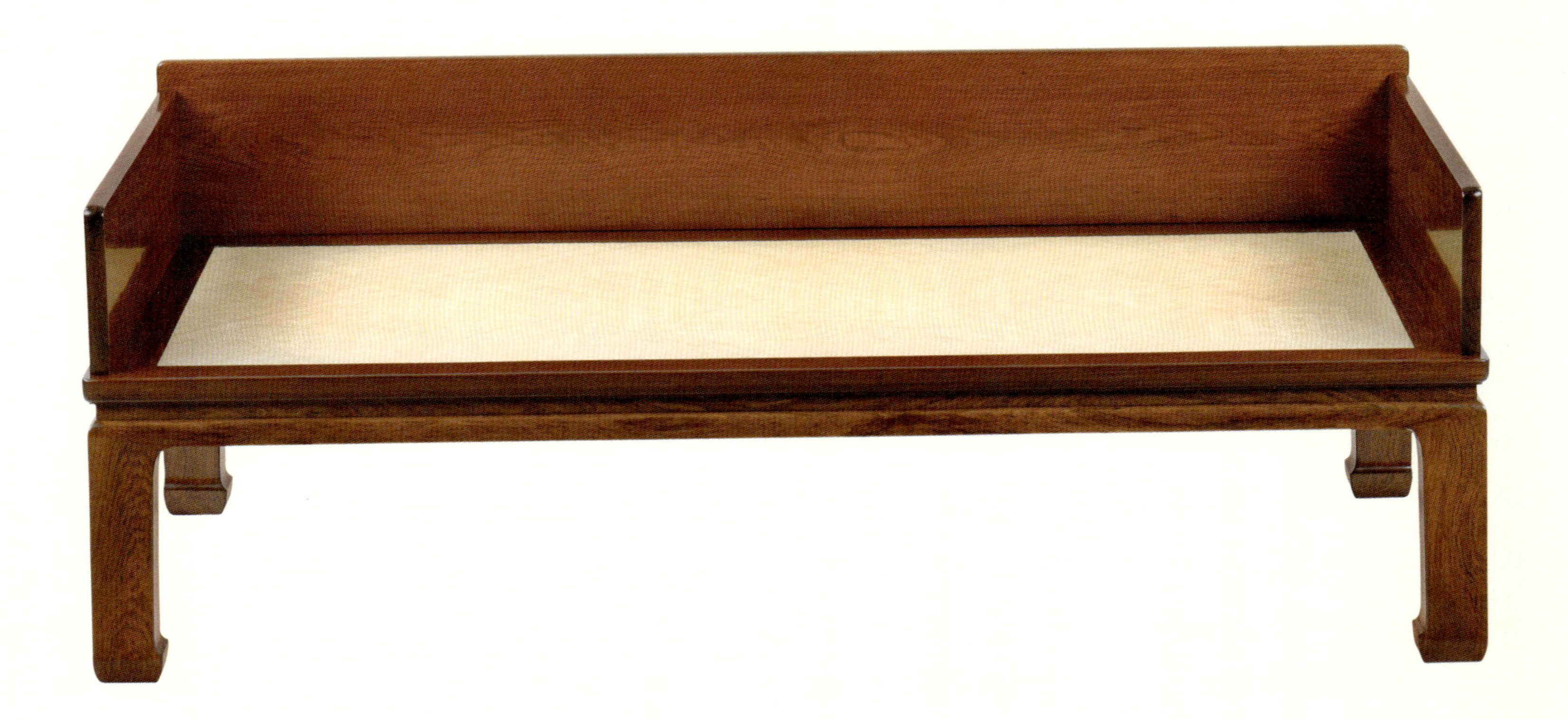

084

独板围子罗汉床（附炕桌）

Arhat Bed with Three One-piece Panels
(with a Tray Table)

206×80×48/82.5cm

罗汉床由白酸枝制成，精选上等好料，充分展示木材的自然纹理。

罗汉床三面围子皆为独板，为了不使纵边的断面木纹露出，同时避免厚独板开裂，在三面独板围子纵边皆拍抹头。正面围子略高，两侧围子略矮，上角皆做罗锅式曲线，正面围子下弯的罗锅曲线高度正好与侧面围子高度一致，以作协调。围子边缘做素混面线脚，饱满圆润。

床面四面攒边，内编席面软屉。边抹做混面，但下面不压边线，只在弧面处理。与常见的束腰式结构不同，座面之下直接接四腿，四腿挺直，下内翻马蹄。两腿之间安修长的直牙条。直牙条窄长，与束腰搭出厚度，才与四腿体量相当，以此增加整件罗汉床的稳定和轻巧。

此件罗汉床造型素雅简洁，马蹄腿直接与床面相接，支撑床体，是其独特之处。

与罗汉床配套的炕桌采用相同设计元素，在桌面之下直接榫接内翻马蹄的四腿，形成统一的造型风格。

The arhat bed is made of white dalbergia cochinchinensis of selected material with gorgeous natural veining.

The bed board is fenced up with one-piece baffle panels on three sides; all are butted with short vertical frames to avoid exposing irregular veining on vertical sections and crackings and fractures; the upper part of central panel is smoothly raised in the middle and curved down a bit at two corners to create a silhouette of hunched back and then naturally connect to lower side panels; edges of three panels are processed in cambered surface.

The bed board is constructed with assembled frames, an embedded soft woven central cushion and simple plain cambered edges. Below the board connects girdled waist of unique style and directly installs four straight legs with inwardly rolled up hoof feet at end; long and narrow apron plates are butted between legs to add a visual horizontal weight to the above girdled waist so as to balance the volumes of four legs and maintain the steadiness and delicacy of the arhat bed.

The bed is built in a style of elegance and implicity; four legs directly butted into the bed board to support the whole structure make it quite unique among others.

The matched tray table which comes along with the arhat bed is built with the same structure: four legs with inwardly rolled up hoof feet at end are directly butted into the table top to maintain a unified style.

085

卍字纹围子八柱架子床

Eight-pillar Canopy Bed with Panels Carved with Buddhist Swastika Tracery

231×192×47.5/231cm

架子床由紫檀制成，色泽沉穆，木纹金星浮现，高贵典雅。

传统架子床一般有四柱和六柱者。四柱者较为少见，三面为围子，一面为门，可供上下；六柱者较为多见，在可供上下的一面增设二柱围门柱，安装短门围子，起到遮蔽使用者头部或脚部的作用，功能更加合理。传统起居方式，架子床多靠墙而置，单面设门，现代的起居方式，则是居中摆放，双面上下，故而此架子床一改传统模式，做成八柱样式，方便使用。

架子床挂檐镶装绦环板，上锼雕为菱花形花格，玲珑剔透，设牙板，沿边起线，至两端微勾卷云纹。八柱除了门柱间外，都设罗锅枨连接，既有力学支撑作用，又可临时搭扶衣物，使用方便。四面围子，上下两层，上层为主，是打洼横竖材攒接而成的卍字纹，又名"步步锦"，寓千秋万代步步锦绣之意；下层装绦环板，浮雕回纹，既作装饰之用，又可遮挡铺设在床上的被褥毯子，取得外观之整齐优美。下座高束腰式，冰盘沿，中间打洼线一道，下方压边线，束腰素直。壶门式牙板，沿边起阳线，腿足三弯，末端内收为卷珠。

此床简繁得宜，装饰自然得体，尚保持明式家具朴质风格。

明崇祯五年
《二刻拍案惊奇》插图
尚友堂刊本

5th Year of Chongzhen Period of Ming Dynasty (1632) Illustration of *Anecdote and Odd Stories*, Shangyoutang's edition

The canopy bed is made of red sandalwood with dark solemn colour and subtle golden pinhole veining (traces of capillary tubes of the plant), revealing a rare luxurious elegance.

Traditional canopy bed is generally constructed with four pillars and six pillars: four-pillar canopy bed is rarly seen and commonly built with baffle panels on three sides, leaving one side open as a doorway allowing for resident to get in and out; six-pillar canopy bed is more commonly seen and has extra two pillars as well as two lower baffle panels installed on the doorway side to functionally shield sleeper's head and feet. According to traditional interior arrangement, the bed is mainly placed against the wall, hence to leave the doorway on one side; however in modern times, the bed is mostly placed in the centre of the bedroom, so resident can get in/out the bed on two sides like this reinvented eight-pillar canopy bed.

The canopy is heavily decorated by hanging eaves embedded with panels delicately carved with lotus floral openwork and apron plates outlined with embossed borderlines and carved with rolling cloud design at corners; between each pillar (except for doorway pillars) installs horizontal hunched rails to serve as mechanical support, as well as clothesline beam; the baffle panels are embedded in two layers: the upper layer is the principal square panel assembled by small wood bricks with embossed borderlines into openwork Buddhist swastika pattern, namely "splendour on each step", meaning wishing for an everlasting splendour; and the lower decorative waist panel carved with bas-relief of fretwork can also practically hide bedclothes to maintain a neat and beautiful appearance; below baffle panels connects a base board with edges narrowed down widthways (also called ice-plate edge) and gouged a groove in the middle, and a rabbet along the bottom rim, and a high girdled waist; below the waist installs *Kun*-gate apron plates with embossed borderlines and three-curved legs (cabriole legs) with foot tips rolled inwardly as pearls.

This canopy bed demonstrates a well-balanced structural design and natural decorative style, still maintaining the simplicity of Ming-styled furniture.

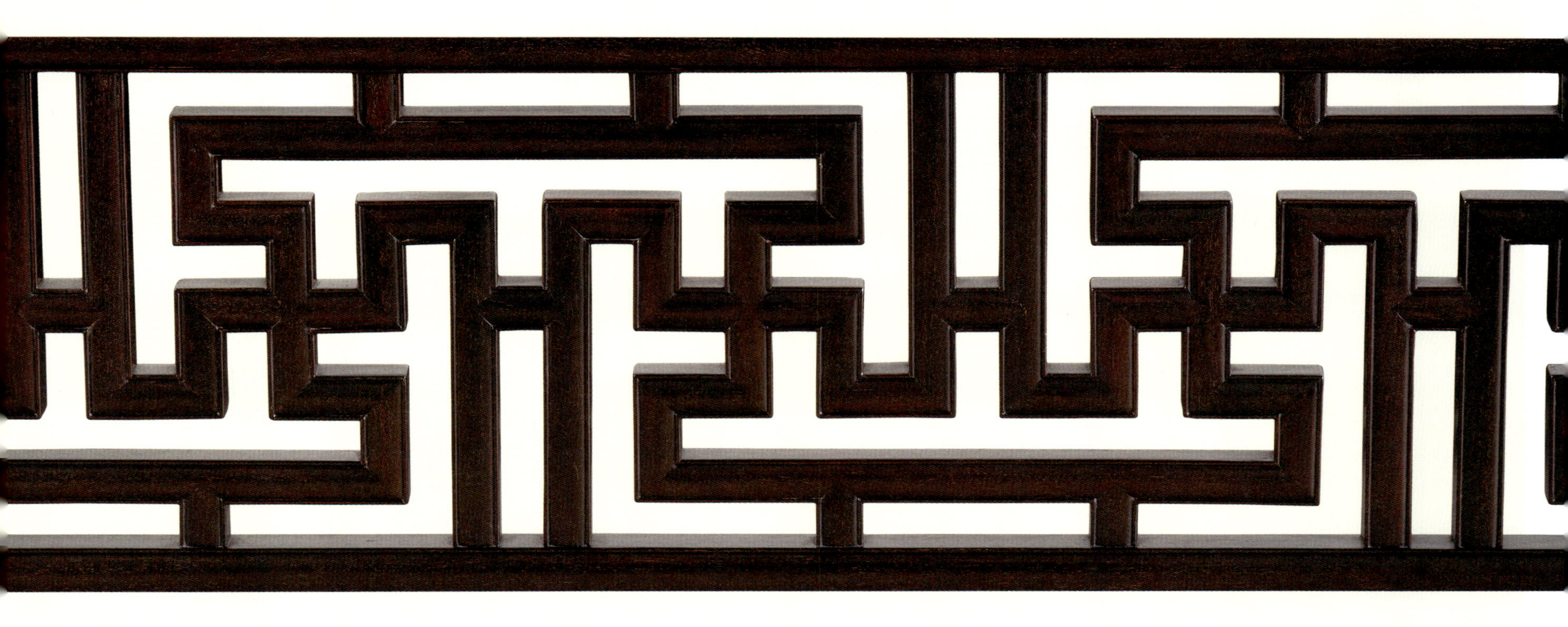

086

竖棂围子八柱架子床

Eight-pillar Canopy Bed with Panels Assembled in Lattice Tracery

231×192×47.5/231cm

架子床由花梨木制成，镶黄杨木雕刻件，颜色搭配加舒朗的造型，清新跳脱，观之心情愉悦。

床顶仰尘攒花格而成，躺卧其上，目可遐游。四面挂檐板，镶装鱼门洞开光的绦环板。前后设门柱，四周围子上下两层，上层较窄，装黄杨木双套环卡子花；下层梳背式，具韵律感，镶嵌有圆光，内嵌黄杨木透雕变体螭龙纹。床座甚为素洁，攒框装板，冰盘沿，直束腰，直牙板，沿边起线，与腿足柔和相交。腿足矩形，下端内翻马蹄腿，典型明式造型，不以雕饰炫目，以精干的造型和比例拿捏打动人心。

竖棂围子即以整齐的竖枨排列而成的围子，因形似梳背，故又俗称为梳背式。据明《工部厂库须知》记载，万历十二年时，宫廷制作家具，“御前传出红壳面揭帖，一本传造龙凤拔步床、一字床、四柱帐架床、梳背坐床各十张”，可见梳背样式在明代晚期宫廷中很流行。

The canopy bed is made of *Huali* wood, embedded with boxwood medallions, giving a visual pleasure owing to its refreshing colour palette and elegantly stretched-out structure.

The canopy is assembled into openwork of geometric flower pattern, leaving a lot to the imagination; below the canopy installs hanging eaves embedded with panels carved with hollowed slim long-fish-doorway design on four sides; doorway posts are installed on the front and back sides; baffle panels on four sides are assembled in two layers: the upper layer is slim rectangular, decorated with boxwood double-circle clamp chips (crocket); the lower layer is assembled in rhythmical comb-back lattice and embedded with boxwood medallions of abstract dragon pattern openwork; the base board is unadorned and clean, constructed with assembled frames with edges narrowed down widthways (also called ice-plate edge), short girdled waist, straight apron plates with embossed borderlines and four legs of rectangular sections with feet curved inwardly as horse hoofs, which is built in typical Ming-style with simple and elegant appearance and perfectly scaled structure.

Baffle panels assembled with neat vertical lattice are also called comb-back lattice due to its resemblance to comb. According to *Account Book of the Ministry of Works* written in the Ming Dynasty, in the 12th year of Wanli Period (1584), regarding to royal furniture making, it keeps a log as"emperor just issued an order written on a red cover official paper to make ten canopy beds of dragon and phoenix design, ten plat beds (couch), ten four-pillar canopy beds and ten beds (couch) with comb-back lattice", which demonstrates the obvious prevalence of comb-back lattice design in the courtyard of late Ming Dynasty.

明 《闵齐伋绘刻西厢记》插图

Ming Dynasty (1368-1644)
Illustration of *Romance of the Western Chamber*,
Illustrated by Min Qiji

榻

Couch

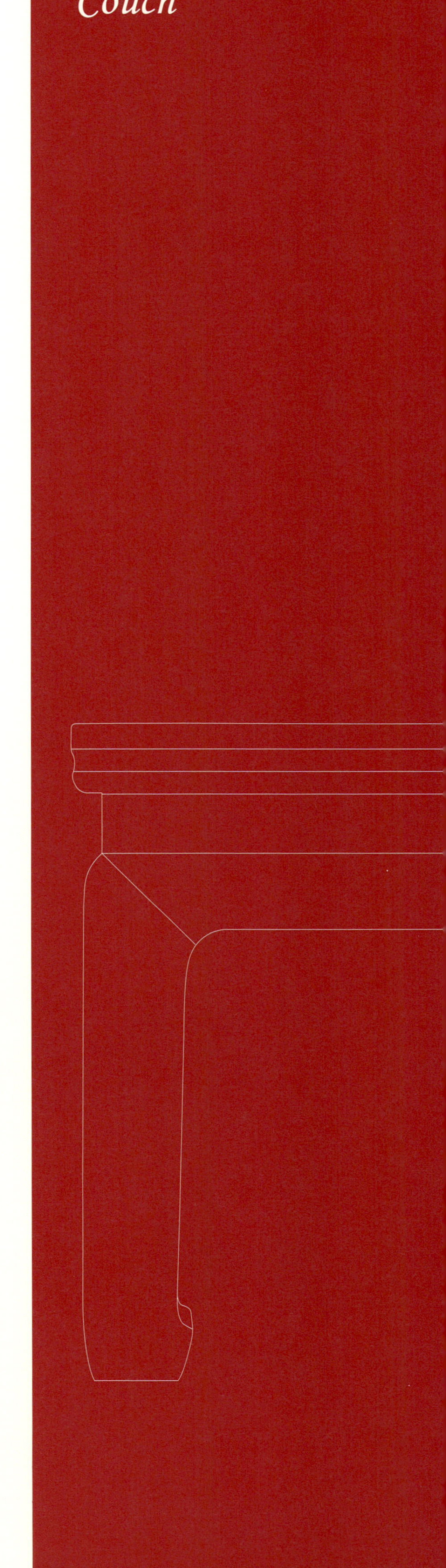

刘熙《释名·释床帐》:"长狭而卑曰榻,言其榻然近地也。小者曰独坐,主人无二,独所坐也。"榻起初是一种矮坐具,有单人的独坐,也有两人并坐者,是较最原始的坐具——席高档一些的坐具,适宜跽坐的起居方式,是席地而坐的低坐系统中最主要的家具之一。唐代时兴一种壸门榻,概自魏晋时已有,宋元时,随着椅子等坐具的普及,榻的使用渐退出主流,但依然存在。明清时期,榻发展成为一种兼具独眠功能的卧具,亦是文房用具之一,是一种没有围子、栏杆的小床,所见传世硬木小榻,造型、做工无不精彩,可惜存量极少。《长物志》云:"古人制几榻,虽长短广狭不齐,置之斋室,必古雅可爱,又坐卧依凭,无不便适。燕衎之暇,以之展经史、阅书画、陈鼎彝、罗肴核、施枕簟,何施不可。"

According to *Definition of Things and Matters* (Chapter of Bed) (written by Liu Xi in the late Eastern Han Dynasty, 25-220), "the furniture with long and narrow top board and short legs are called *Ta* (couch), meaning close to the ground; and the small couch is for one person to rest on." *Ta* (couch), originally a low-seat furniture, used by single person or two persons, is more advanced than the primitive seating device—mat, and is for person to kneel on, as one of the major furniture in the low-seat system. In the Tang Dynasty (618-907), a widely pursued couch with *Kun*-gate design actually appeared as early as Kingdom of Wei (220-265) and Jin Dynasty (265-420); in the Song Dynasty (960-1279) and Yuan Dynasty (1206-1368), with the widespread availability of chairs, *Ta* (couch) gradually receded to the background of the history. In the Ming and Qing Dynasties, couch was used as recliner for one person to sleep on, and also a common item in the study of literati: it served as small bed without baffle panels or rails. The exsiting small couches made of hard wood are built with exquisite structure and high-skilled craftsmanship, and extremely rare. According to *Superfluous Things* written by Wen Zhenheng in the late Ming Dynasty, "the ancient end table (*Ji*) and couch (*Ta*), though varied in length and width, are of special archaic beauty while being placed in the meditation chamber, and can be used as bed, couch or armrest whenever the owner wanted to read books, to appreciate paintings, to display bronze antiques, to serve snacks, or to make the bed."

087

高束腰马蹄腿榻

Couch with High Girdled Waist and Hoof Feet

206×86×48cm

榻由铁梨木制成，色泽沉穆含蓄，颇耐品鉴。

高束腰式，即束腰的高度大于座面冰盘沿厚度。束腰结构是从古代建筑须弥座形式衍化而成，高束腰则是这种形式较早期的样式。

榻座面攒框，编织席面为软屉，座面冰盘沿造型醇厚，上方平齐，下方兜进如唇口，只在中间起一道锐利的洼线，丰富层次变化。束腰平素，腿足上部露明，但与束腰平齐，仍然保持平素的造型特点。牙板微鼓，溜肩，弧线拿捏甚好，柔中寓刚。直牙条，沿边起碗口线。所谓碗口线，即边缘柔和翻起，如同古代瓷碗的边缘，是中国家具中制作难度较大、造型效果极雅的一种线脚。直腿足，下方内翻马蹄足，大小适宜，端庄挺拔。

此榻周身不见雕饰，但处处精工细作，将明式家具含蓄典雅的风骨展现无疑。搁置客厅、书房、卧室，或坐或卧，或搁置物件，无所不可。

The couch is made of lignumvitae with dark and solemn colour.

The high girdled waist, meaning the height of the waist is higher than the width of seat panel's edge which is narrowed down widthways (also called ice-plate edge). Such structure is derived from Sumeru seat in ancient architecture and the high girdled waist is the early form of such design.

The seat panel is construced with assembled frame and a soft woven cushion; seat edges are narrowed down widthways (also called ice-plate edge) and gouged out a groove in the middle to add more variations; the girdled waist is unadorned and to connect to four legs directly; straight apron plates with embossed borderlines bulge out slightly to form a smooth curvy silhouette, balancing softness and hardness; the embossed borderline, also called bowl-rim borderline, locks the plates with soft rounding rims bearing the resemblance to the ancient porcelain bowl rim, as an extremely elegant and hard-to-process moulding in Chinese furniture; below it installs four straight legs with feet curved inwardly as horse hoofs. The general structure is well proportioned and of upright appearance.

The couch is unadorned, yet processed with delicacy and care, demonstrating the elegance of Ming-style furniture; it can be placed in the living room, study, or bedroom, can be seated or reclined in, or used as display platform.

明万历
《古杂剧二十种》插图
顾曲斋刻本

Wanli Period of Ming Dynasty (1573-1620)
Illustration of *Twenty Classic Chinese Operas*, Guquzhai's edition

088

三弯腿卷珠足榻

Couch with Three-curved Legs (Cabriole Legs) and Rolled-pearl Feet

206×70×49cm

榻以红酸枝制成，色泽沉稳内敛，攒框装席面为软屉，浅色座面与周匝深色边框形成对比，相得益彰。

此榻属于造型甚为优雅的一种，三弯腿足甚是惹人注目。边框冰盘沿，中间打洼线一道，下方有压边线，直素束腰，壶门牙板，曲线柔婉流畅，与腿足走势联为一气，形成和谐统一的整体造型。牙板沿边起圆润的阳线，至中间勾卷相交，往两端则衍化为相背的卷云纹。三弯腿，内翻花叶状造型，亦饰勾云纹，下端外翻卷珠，下踩圆足。整体观之，三弯腿足又如同大象卷鼻抽象而来，寓意吉祥。

明万历 《列女传》插图
鲍氏知不足斋刊本

Wanli Period of Ming Dynasty (1573-1620)
Illustration of *Biography of Famous Women in Ancient China*, Zhibuzuzhai's edition from Family Bao

The couch (*Ta*) is made of red dalbergia cochinchinensis with dark and subtle colour; the seat panel is construced with dark reddish brown assembled frame and a beige soft woven cushion, forming a strong colour contrast.

This couch is built in elegant structure, with striking three-curved legs (cabriole legs), neat panel edge narrowed down widthways (also called ice-plate edge) with a clear groove gouged out in the middle and a rabbet along the bottom rim, an unadorned girdled waist, *Kun*-gate apron plates with smooth curves connecting four cabriole legs to give a unified style; apron plates are decorated with embossed borderlines which cross in the centre and twist as floral knot and roll up as cloud pattern at corners; cabriole legs, curve inwardly as floral leaves decorated with cloud pattern in the middle, and curve outwardly at end to form rolled-pearl feet with four round foot pads attached below. On the whole, these three-curvd legs bear close resemblance to trunks of elephants (*Xiang*), indicating "auspicious"(*Ji Xiang*) due to the homophones in Chinese.

庋具

Furniture-to-store

庋者，庋藏也，庋具是收纳、收藏物品的器具，显然，它是文明发展到一定程度的产物，有所余，方需藏。早期的庋具主要是柜和箱，其时柜的造型与今日的柜相差较多，更像是带足的大箱子，上方开孔为柜门，有点像北方一直在使用的躺柜，唐三彩柜模型出土数例，均是如此。箱自诞生至今，虽有发展，但无大的变化，主要有盝顶和平顶两种，规格则既有可容数人的大躺箱，也有精巧可把玩的小百宝箱。橱是后来出现的庋具，本来与柜的不同之处在于前方设门，所以大多是竖高而进深短的庋具。随着家具的发展，橱和柜的概念逐渐混淆，前方设门的庋具都逐渐成为橱柜。总体来说，柜的概念所指更多，橱开始主要专指联二橱、闷户橱这些兼备承具和庋具功能的家具。另外本书将架格、柜格也纳入庋具，因为他们的造型与橱柜无二，只是不全设柜门而已。

Furniture-to-store, is used to store and to collect things. Apparently, it's the result of a highly developed civilization in which extra wealth and production lead to extensive demand for storing and collecting. Early furniture used to store and collect properties are mainly cabinet and chest: early cabinet is more like today's big lying chest widely used in the North of China with legs and the top panel with holes bored through as lid, such as several unearthed tri-colour glazed miniature pottery cabinets. The chest, since its birth, went through few changes through history, and has mainly two types: one with truncated pyramid top and the other with flat top; regarding to the size, it includes big lying chest which can hold several adults in it and also small treasure chest which can be played in palms. The cupboard (coffer) appeared afterwards; originally, it has front doors instead of top panel doors like the early cabinet, therefore, early cupboard (coffer) was tall and with shallow chamber; with the development of ancient furniture, concepts of cupboard and cabinet were gradually mixed up, more storing furniture with front doors were then categorized into the cabinet; generally, the cabinet has a broader meaning, whereas, cupboard (coffer) started to specifically indicate two drawer coffer and coffer with drawers and a hidden compartment, which can both function as display furniture and storing furniture. Furthermore, this book also includes shelf, stand, and display cabinet into the chapter of furniture-to-store, as their forms and structures are almost same with the cabinet except for the setting of doors.

橱

Cupboard

橱亦写作“厨”，在古代本指前有柜门的庋具，但是从家具史研究的角度，这类庋具都纳入柜类，是橱柜的广义所指。而这里狭义的橱，则特指联二橱、联三橱或闷户橱这类兼备承具和庋具的特殊家具，这类橱在今日多被搁置玄关，既实用，又美观。所谓联二、联三，多指一种案形结体的家具，多有翘头，使得整体造型不失于笨重，腿足外侧装角牙，内侧上方装抽屉，两具并列者名联二，三具则名联三；下方多为闷仓，故又名闷户橱。此外，还有一种下面设柜门者或四面平结体者。橱的装饰繁简不一，简练者多在抽屉面上外贴或浮雕一层壸门券口，装饰的同时也方便安装铜饰件，两侧和下方牙板做成光素刀牙板，复杂者则将挂牙加长，并在挂牙和下方牙板上透雕螭龙、螭凤、卷草等图案。

The cupboard (橱), also written as " 厨 "in Chinese, originally indicating storing furniture with front doors, however was later categorized into cabinet as " 橱柜 ". Today's narrowed meaning of cupboard specifically indidates two-drawer coffer, three-drawer coffer, or coffer with drawers and a hidden compartment, which can both function as display furniture and storing furniture, which are mainly placed at the hallway or foyer, with practical and aesthetic values. The two/three-drawer coffer is built in narrow table structure, mostly with everted flanges to give it a sense of lightness, with corner plates butted along the side of legs, and with drawers installed between legs, hence the name two-drawer or three-drawer coffer; below the drawers, there usually has a hidden compartment. Moreover, some coffers have lower doors or flat-corner structure. Decorations of cupboard are also varied; the simple ones will paste or carve a frieze of relief on the drawers as ornament and the base for metal fittings, and leave apron plates unadorned; the complicated ones will extend the hanging plates with openwork of dragon, phoenix and floral design.

089

卷草纹联二橱

Double-decked Two-drawer Cupboard Carved with Arabesque Bas-relief

133×45×82cm

联二橱是闷户橱的一种，此闷户橱由黄花梨制成，选料精良，纹理流畅自然，麦穗纹满密。

门户橱面四面攒边而成，面心平装薄板。边抹做冰盘沿压边线线脚处理。抹头做翘头，翘头曲线柔婉舒展。

橱面下留出吊头，安四腿，腿截面为混面，内边缘压边线线脚处理。四腿八挓，收分明显，以增加稳定感。吊头下安角牙，角牙上雕刻云头卷草纹饰。

橱面下两腿间安两抽屉，抽屉以短柱界开，抽屉下安横枨。抽屉脸贴壶门券口牙子，正中镶铜制圆面叶，安环形拉手，并设锁鼻和锁销。抽屉之下横枨之下再安横枨，两横枨之间打槽装板，做闷仓。闷仓精选独板黄花梨，光素无饰，以显示出黄花梨行云流水的优美纹理。下横枨之下安花牙子，牙条做壶门曲线，花心部分向下坠出云头牙子，向上翻转成浅浮雕卷草纹。两侧向上翻转，与牙头相接，翻出云头纹。牙子边缘起阳线，以勾勒轮廓。

此联二橱为典型的闷户橱造型，翘头起翘精神，加之适当雕琢装饰，整件家具显得稳重精巧。

Two-drawer cupboard (coffer) is one kind belonging to the category of cupboard with a hidden compartment. This cupboard is made of yellow *Huali (Huanghuali)* wood of refined material with smooth and dense natural veining.

The top board is constructed with a flat central panel and assembled frames with edges narrowed down widthways (also called ice-plate edge) and planed down to a clean rabbet along the bottom rim as borderline; two short edges are butted with everted flanges of beautiful curvy silhouette; four legs with cambered surface and a rabbet running along the inner edges are installed indentedly to leave space for hanging corner plates decorated with carvings of cloud and floral design on two sides below the top board. Four splayed legs form a steady structure.

Between legs installs two drawers divided by a short strut and a stretcher below; each drawer's front surface is pasted with *Kun*-gate apron plate and inlaid with copper-made medallion handle base strung with demi-circle handle and also installed an escutcheon and a lock pin. Below drawers and the stretcher installs an extra lower stretcher to brace additional side panels in between, forming a hidden compartment made of one-piece yellow *Huali (Huanghuali)* wood of refined material without any decorations to bring out its graceful natural veinings. Below the lower stretcher installs *Kun*-gate apron plates carved with bas-relief of floral pattern in the middle and clould design at corners, and outlined with embossed borderlines.

This two-drawer cupboard is built with typical structure of cupboard with a hidden compartment; two everted flanges are vivid and full of spirit; with appropriate decorations and carvings, the whold piece appears with solemnity and lightness.

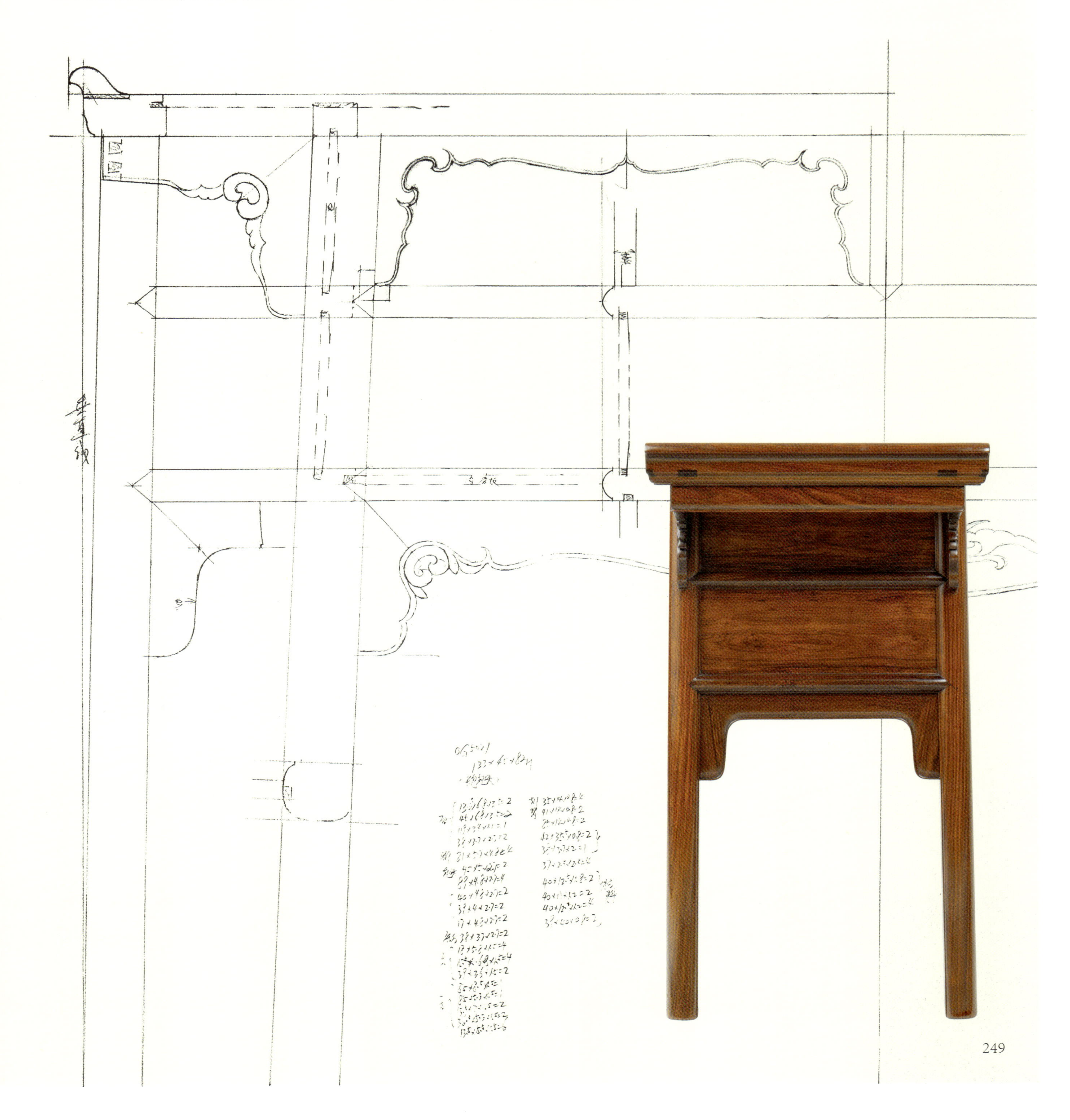

090

联三橱

Double-decked Three-drawer Cupboard Carved with Arabesque Bas-relief

196×50×86cm

此联三橱由白酸枝制成，选料精良，色泽淡雅莹润，纹理细腻优雅。

橱面四面攒边，面心平装薄板，边抹做冰盘沿压边线线脚处理。抹头做翘头，翘头曲线温婉舒展。

橱面下留出吊头，装四腿，吊头下安素角牙，只在边缘起阳线。四腿八挓，收分明显，以增加稳定感。腿截面为混面，内侧边缘压边线线脚处理。橱面下两腿间安两横枨，界成双层，上层以两短柱界出三抽屉，下层闷仓以一短柱界出两段，内打槽装板，装纹理优美的独板。抽屉前脸贴壸门券口牙子，脸正中镶铜制圆形面叶，面叶上安环形拉手，并设锁鼻和锁销。下横枨与两腿之间安刀板牙子，牙子光素，只在边缘起阳线，与吊头处的角牙相呼应。

此联三橱选料精良，工艺精巧，造型简洁，起翘精神。

This cupboard is made of white dalbergia cochinchinensis of refined material with smooth veining and elegant gloss.

The top board is constructed with a flat central panel and assembled frames with edges narrowed down widthways (also called ice-plate edge) and planed down to a clean rabbet along the bottom rim as borderline; two short edges are butted with everted flanges of elegant curvy silhouette.

Four legs with cambered surface and a rabbet running along the inner edges are installed indentedly to leave space for hanging corner plates on two sides below the top board. Four splayed legs form a steady structure. Between legs installs two stretchers to divide the front side into two decks: upper part holds three drawers divided by two short struts; lower part holds a hidden compartment divided by a short sturt. The whole body structure is embedded with one-piece material of beautiful veining; each drawer's front surface is pasted with *Kun*-gate apron plate and inlaid with copper-made medallion handle base strung with demi-circle handle and also installed an escutcheon and a lock pin. Below drawers and stretchers installs unadorned apron plates outlined with embossed borderlines to echo the plain hanging corner plates on two sides below the top board.

This three-drawer cupboard is made of highly selected material and built with skillful craftsmanship, demonstrating the beauty of simplicity and liveliness.

091

螭龙纹联三橱

Double-decked Three-drawer Cupboard Carved with Openwork of Dragons

196×50×86.5cm

此联三橱由黄花梨制成，色泽温润。橱面四面攒边，面心平装薄板，边抹做冰盘沿压边线线脚处理。边抹做翘头，翘头曲线优雅温婉。

橱面下留出吊头，安四腿，四腿八挓，收分明显，以增加稳定感。吊头与腿足连接处安挂牙，角牙透雕螭龙纹饰。橱面下、腿足间安两横枨三短柱，界出上层三抽屉，下层闷仓两段装板。抽屉脸贴壶门券口牙子，牙子之上浅浮雕卷草纹。抽屉脸正中镶方形面叶，上安环形拉手，并设锁鼻和锁销。

下横枨之下安壶门牙子，由一块大板制成，勾勒出壶门曲线，壶门曲线正中向下卷出三朵灵芝纹，壶门曲线两侧之下透雕螭龙纹。

此件联三橱在标准闷户橱基础之上，吊头之下的挂牙和下面的牙条之上均透雕螭龙纹，华美富丽，装饰性很强。

This cupboard is made of yellow *Huali (Huanghuali)* wood of refined material with graceful gloss. The top board is constructed with a flat central panel and assembled frames with edges narrowed down widthways (also called ice-plate edge) and planed down to a clean rabbet along the bottom rim as borderline; two short edges are butted with everted flanges of elegant curvy silhouette.

Four legs are installed indentedly to leave space for hanging corner plates carved with openwork of dragon design on two sides below the top board. Four splayed legs form a steady structure. Between legs installs two stretchers to divide the front side into two decks: upper part holds three drawers divided by two short struts; lower part holds a hidden compartment divided by a short sturt; each drawer's front surface is pasted with *Kun*-gate apron plate carved with bas-relief of floral patterns and inlaid with copper-made sqaure handle base strung with demi-circle handle and also installed an escutcheon and a lock pin.

Below drawers and stretchers installs *Kun*-gate apron plates made of an one-piece material, and outlined with embossed borderlines, and carved with openwork of three ganodermas in the middle and openwork of dragon design at corners.

This three-drawer cupboard is based on the basic structure of cupboard with a hidden compartment, with additional opulent openwork of dragon design applied on hanging corner plates, showcasing the ultimate luxurious magnificence.

092

螭龙纹联三橱

Double-decked Three-drawer Cupboard Carved with Openwork of Dragons

196×50×86.5cm

此联三橱由紫檀制成，紫檀色泽沉穆，纹理隐现，牛毛纹满密。

此联三橱橱面四面攒边，面心平装薄板，边抹做冰盘沿压边线线脚处理。抹头做翘头，翘头曲线柔婉舒展。

橱面下留出吊头，安四腿，四腿八挓，收分明显，以增加稳定性。吊头下腿外侧安挂牙，挂牙较大，满雕螭龙纹饰。橱面下、两腿间安两横枨三短柱，在上层界出三个抽屉，下层闷仓界出两段装板。抽屉脸贴券口牙子，正中镶铜制方形面叶，装环形拉手，还设锁鼻和锁销。铜制面叶金属光泽浅淡，与色泽深穆的紫檀形成鲜明对比。

下横枨之下安牙子，牙子勾勒出壸门曲线，壸门曲线正中向下翻出三朵灵芝纹，壸门曲线两侧之下透雕螭龙纹。

此联三橱造型经典，在简洁精炼的构件基础之上，在吊头下的挂牙和横枨下的牙子上透雕大块的螭龙纹饰，为此件家具增加了华美富丽的气质。

This cupboard is made of red sandalwood with dark and solemn colour and subtle ox-hair veining (traces of capillary tubes of the plant).

The top board is constructed with a flat central panel and assembled frames with edges narrowed down widthways (also called ice-plate edge) and planed down to a clean rabbet along the bottom rim as borderline; two short edges are butted with everted flanges of elegant curvy silhouette.

Four legs are installed indentedly to leave space for large hanging corner plates carved with openwork of dragon design on two sides below the top board. Four splayed legs form a steady structure. Between legs installs two stretchers to divide the front side into two decks: upper part holds three drawers divided by two short struts; lower part holds a hidden compartment divided by a short sturt; each drawer's front surface is pasted with *Kun*-gate apron plate carved with bas-relief of floral patterns and inlaid with copper-made sqaure handle base strung with demi-circle handle and also installed an escutcheon and a lock pin. The light metallic lustre of the copper fittings forms a stark contrast with the dark colour of red sandalwood.

Below drawers and stretchers installs *Kun*-gate apron plates outlined with embossed borderlines, and carved with openwork of three ganodermas in the middle and openwork of dragon design at corners.

This three-drawer cupboard is built in classic structure decorated with large opulent openwork of dragon design on hanging corner plates and lower apron plates, enhancing a sense of luxurious magnificence.

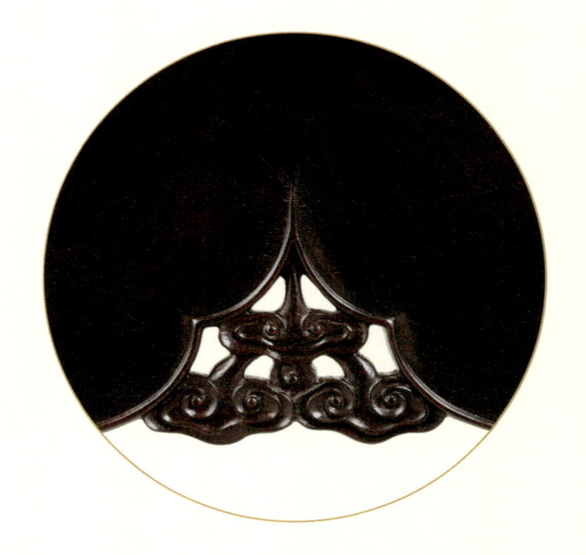

093

卷云纹联三橱

Double-decked Three-drawer Cupboard
with Cloud-shaped Apron Plates

176×48×85cm

此柜橱由白酸枝制成，白酸枝色泽淡雅莹润，纹理流畅自然。

橱面四面攒边，面心平装薄板，边抹做冰盘沿压边线线脚处理。抹头做翘头，翘头曲线柔婉优雅。

橱面之下留出吊头，装腿，四腿八挓，收分明显，以增加稳定感。吊头之下、腿外侧安云头角牙。橱面下、两腿间安横枨，横枨之上以两短柱界出三抽屉，抽屉脸光素，正中镶铜制圆面叶，面叶上安圆形拉环。横枨之下，靠近腿下端再安横枨，两横枨之间正中安竖柱，竖柱两侧各安两个四面攒边的门板。两个门板之间各安两圆形合叶。竖柱上安较大的圆形面叶，上安吊牌和锁鼻。

下横枨之下安刀板牙子，牙条长直光素，牙板两侧向下起弧线与牙头连接。

此柜橱体量较大，稳重大气，比例协调，造型简练。

This cupboard is made of white dalbergia cochinchinensis of refined material with smooth veining and elegant gloss.

The top board is constructed with a flat central panel and assembled frames with edges narrowed down widthways (also called ice-plate edge) and planed down to a clean rabbet along the bottom rim as borderline; two short edges are butted with everted flanges of elegant curvy silhouette.

Four legs are installed indentedly to leave space for hanging corner plates with carvings of cloud design on two sides below the top board. Four splayed legs form a steady structure. Between legs installs three drawers divided by two short struts and a stretcher below; each drawer's front surface is inlaid with copper-made medallion handle base strung with demi-circle handle. Below drawers and the stretcher installs an extra lower stretcher to brace a vertical post butted in the middle; on two sides of the post connects four hinged door panels constructed with assembled frames; and the vertical post is installed with a large medallion lock base, as well as an escutcheon and a lock pin.

Below the lower stretcher installs unadorned apron plates with rounded corner plates.

This cupboard is built in large size with balanced scale and simple structure, giving a sense of solemnity and grandeur.

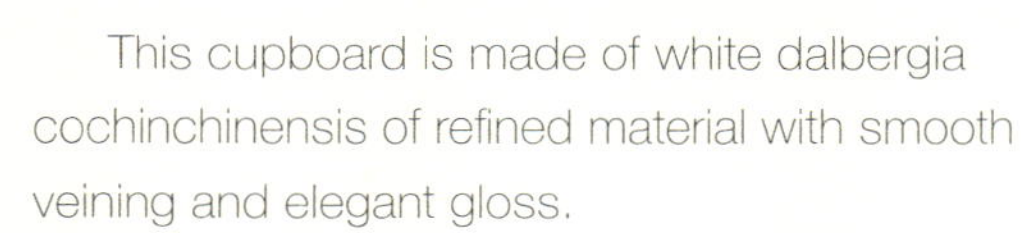

格

Display Cabinet

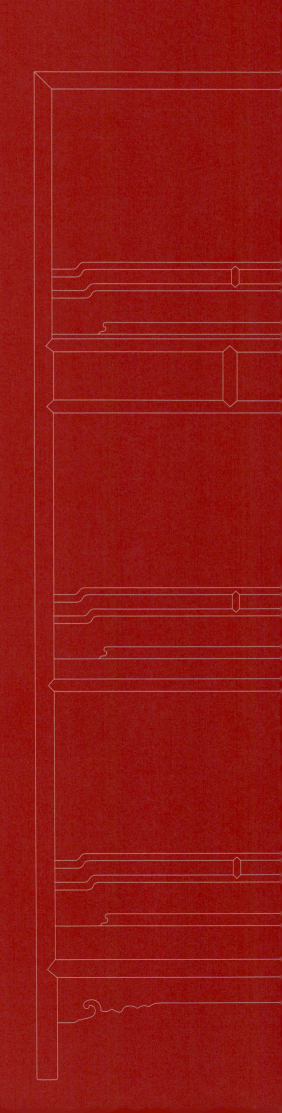

格本亦属架具，但其造型与橱柜相近，多设抽屉、柜门，兼具庋具的功能，故这里将之纳入庋具。就其形式来说，有架格和柜格之分。架格多称为书架，多三层，设搁板，简练者或四面空敞，或三面设素板封堵。较复杂者在每层的后、左、右三面设围栏，有的做成栏杆样式，舒朗可爱；有的透雕各式花纹，富于装饰。在中间一层处，多并设两具抽屉，可以容纳零碎物件，实用的同时还增加层次变化。柜格者，是一种下面设对开门、上面两层或三层搁板的格，也有少见的中间设柜门，上下为搁板者。架格和柜格在书房、客厅皆可摆放，柜和抽屉收纳物件，敞开的格内可放置书册，也可摆放古董等物品，如同展示柜，以供欣赏，是书卷气甚浓的一类家具。

The display cabinet, basically functions as shelf; yet due to its similar appearance with cabinet, as well as its additional structures of drawers and doors which allow it to serve as storing furniture, the display cabinet is categorized as furniture-to-store in this book. In terms of its form, it can be divided into display shelf and display cabinet: the display shelf, commonly called book shelf, mostly has three storeys separated by two shelves, left its four sides exposed or three sides sealed with plain boards, or installed balustrade-styled fenders on the back, left and right sides of each storey (sometimes decorated with openwork of various designs and patterns); there often adds two drawers in the middle storey to store small things and items to further its practical value as well as to increase spatial variation; the display cabinet often has doors on the bottom storey, leaving upper two or three storeys exposed; in a few cases, there adds doors in the middle storey, leaving the upper and the lower storeys exposed. The display shelf and display cabinet are usually placed in the study or living room. The cabinet and drawers can store small items; and the exposed shelf can hold books or antiques as display case for people to appreciate, therefore, they are typical furniture for literati's room.

094

四面平三层架格

Three-Tier Display Shelf with Flat-corner Structure
(Corners Butted with Mitered Corner Bridle Joints)

108×41.7×190.5cm

此架格由花梨木制成，花梨木色泽淡雅莹润，纹理优美流畅。

架格皆用方材，用料纤细。在四腿顶部加横枨和侧面枨，以综角榫连接，形成四面平结构。横枨之下安更细横枨，四面交圈，横枨之间承架格板，形成一层，可以摆放卷轴、书籍之类。往下再安两横枨，横枨之间以短柱界出三抽屉，抽屉脸正中装铜制吊牌，如此形成第二层。腿足偏下位置再安横枨，横枨四面交圈，承架格板，如此形成第三层。此横枨之下，安刀板牙子，牙子光素无饰。

此架格造型简素，用料单细，抽屉亦薄，给人轻巧舒朗的视觉感受。

This display shelf is made of *Huali* wood with elegant colour, smooth and natural veining.

It's assembled with long and slim components of square section. Between top ends of legs installs stretchers butted with mitered corner bridle joints to form a steady flat-corner structure; each shelf panel in the middle is braced by four slimmer stretchers to hold scrolls and books; below the middle storey installs an extra stretcher to separate the room for three drawers divided by two short struts; each drawer's front surface is inlaid with copper-made flower-shaped handle base strung with H-shaped handle. Between lower ends of legs installs the bottom shelf panel braced by stretchers; below the stretchers are butted with unadorned apron plates.

This display shelf is built in simple structure with same slim square-sectioned material to give an ethereal and elegant visual effect.

095

罗锅枨栏杆三层架格

Triple-Tier Display Shelf with Flat-corner Structure (Corners Butted with Mitered Corner Bridle Joints) and Hunched Rails

108×40×196cm

此架格由紫檀制成，紫檀色泽深穆，纹理细腻，牛毛纹满密。

架格皆用方材，用料纤细。在四腿顶部加横枨和侧面枨，横枨、侧面枨和四腿以综角榫连接，形成四面平结构。横枨之下，安双横枨，两枨之间以短柱界出三抽屉，抽屉脸光素，安铜制吊牌。上面横枨之上安罗锅枨栏杆，栏杆上安双罗锅枨，罗锅枨之间安两矮老，形成错落空间。抽屉之下再安两层横枨和栏杆。最下层横枨之下安云头牙子，云头牙子曲线翻转自如，边缘不起阳线。

此架格以横枨、侧面枨、架格板和栏杆组成三层，可摆放卷轴、书籍等文房之物。此架格各构件用料纤细，抽屉亦薄，栏杆做双罗锅枨加矮老的做法极罕见，勾勒了独特的栏杆空间。

This display shelf is made of red sandalwood with dark and solemn colour and subtle ox-hair veining (traces of capillary tubes of the plant).

It's assembled with long and slim components of square section. Between top ends of legs installs stretchers butted with mitered corner bridle joints to form a steady flat-corner structure; each shelf panel in the middle is braced by four stretchers; below the first storey installs an extra stretcher to separate the room for three drawers divided by two short struts; each drawer's front surface is inlaid with copper-made flower-shaped handle base strung with H-shaped handle; above each shelf panel installs balustrade-styled fenders decorated with double hunched rails (fixed with two short struts in the middle) on the back, left and right sides to add spacial variation; below the bottom shelf panel installs apron plates carved with abstract cloud design of curvy silhouette.

This display shelf has its shelf panels braced by stretchers between legs to separate three storeys. It can hold scrolls, books and antiques. The shelf is built with slim square-sectioned material, has shallow drawers and balustrade-styled fenders uniquely decorated with double hunched rails to frame each storey as a beautiful scene.

096

攒后背水波纹两层架格

Double-Tier Display Shelf with Assembled Back Panel of Ripple-motif

98×45×168cm

此架格由白酸枝制成，精心选料，色泽淡雅莹润，纹理流畅自如。

架格各构件皆为圆材，为圆角柜造型，格顶四面攒边，边抹做泥鳅背压双边线线脚处理。格顶略出帽，安装四圆腿，四腿截面外圆内方。四腿八挓，收分明显，以增加稳定性。

四腿中间位置，安两横枨，两横枨之间以短柱界出两抽屉，抽屉脸光素，正中铜制吊牌。格顶之下、上横枨之间安壶门券口牙子，牙子边缘起阳线，以勾勒轮廓。四腿偏下位置再安横枨，置架格板，形成第二层。横枨之下安牙子，牙子两端翻出卷草，牙子边缘起阳线，以勾勒轮廓。抽屉之下、下横枨之上亦安壶门券口牙子，牙子边缘起阳线。

架格后背攒斗出水波纹锦地，波纹相接触的部分浅浮雕四瓣朵花。

此架格以圆材构架，抽屉两具，两侧，每层各设三面壶门式券口，后背攒斗出水波纹锦地，造型独特，气质雅致。

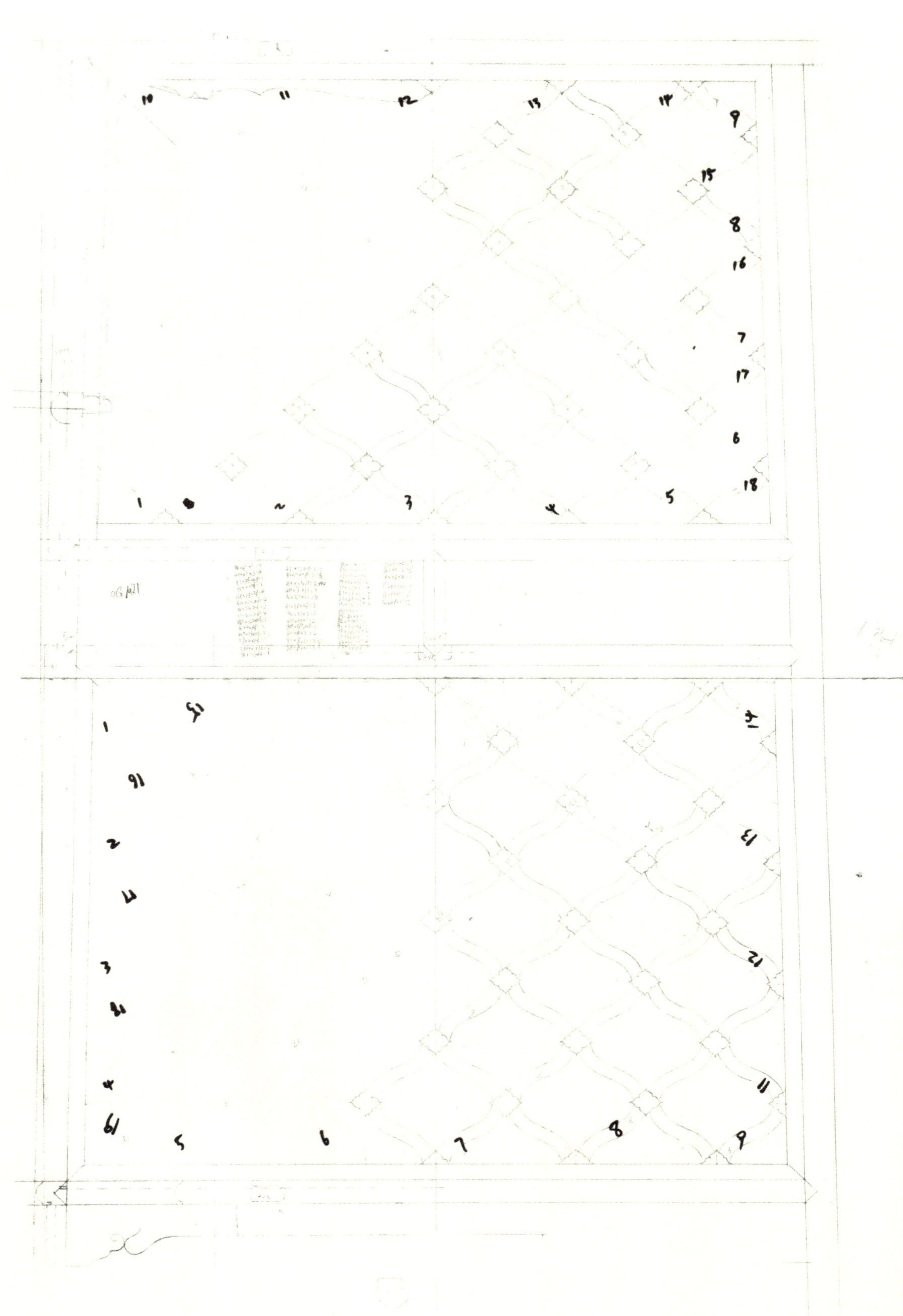

This display shelf is made of white dalbergia cochinchinensis of refined material with smooth veining and elegant gloss.

It's assembled with round-sectioned components and built in the structure of round-cornered cabinet: the top panel is braced with assembled frames with edges polished in loach-back style and planed down two rabbets on both sides as borderlines; below the top panel installs four splayed legs of half-round-and-half-square section to increase the steadiness of the structure.

Between legs installs two stretchers to separate the room for two drawers divided by one short strut; each drawer's front surface is inlaid with copper-made flower-shaped handle base strung with H-shaped handle; below the top panel is butted with arabesque apron plates of *Kun*-gate style outlined with embossed borderlines; below the bottom shelf panel installs apron plates carved with arabesque design and outlined with embossed borderlines; below the drawers also installs arabesque apron plates of *Kun*-gate style outlined with embossed borderlines.

The display shelf is sealed with an assembled back panel of water-wave-shaped latticework decorated by bas-relief of small flower crockets at each point of intersection.

This display is constructed with components of round section, with two drawers and arabesque apron plates of *Kun*-gate style butted on three sides in each storey, and an assembled water-wave-shaped latticework of the back panel, truly revealing a unique characteristics of elegance beauty.

097

罗锅枨栏杆三层柜格

Triple-Tier Display Cabinet with Hunched Rails

100×40×203cm

柜格由白酸枝制成，精心选料，色泽淡雅莹润，纹理流畅自然。

柜格后背空敞，三面设罗锅枨栏杆，第三层之下平列抽屉两具，以短柱界出。抽屉脸光素，正中安铜制吊牌。

两抽屉之下安两门，门板四面攒边，中装薄板。门板与柜腿以条形委角合叶连接，中间安条形委角面叶，安吊牌与钮头，为硬挤门做法。柜门之下安横枨，横枨之下安云头牙子，牙子光素，不起阳线，以突出亮格柜主体。

此柜格上面开敞，下面安抽屉和柜，形成上虚下实的柜体空间。上面开敞的三层使用罗锅枨栏杆，很是独特，将开敞空间界出空间感，又通透空灵，意趣无穷。

This display cabinet is made of white dalbergia cochinchinensis of refined material with smooth veining and elegant gloss.

The upper three-storey shelf of the display cabinet is installed with balustrade-styled fenders decorated with double hunched rails (fixed with two short struts in the middle) on the back, left and right sides of each storey, and left with its back side exposed; below the bottom storey installs two drawers divided by a short strut; each drawer's front surface is inlaid with copper-made flower-shaped handle base strung with begonia-shaped handle.

Below the drawers installs two doors constructed with assembled frames and central door panels; doors are connected to legs by rectangular hinges with dented-corners, and inlaid with rectangular handle base strung with begonia-shaped handle and pins; below the doors installs a stretcher and apron plates carved with cloud design.

This display cabinet is exposed in upper three storeys and has two drawers and one cabinet below to form a contrasting spacial variation; the unique alustrade-styled fenders decorated with double hunched rails help to divide the space and present a celestial and lively sense of beauty.

柜

Cabinet

柜原作“匮”。河南灵宝张湾汉墓曾出土一个陶柜模型，而山东沂南汉画像石上也出现了躺柜的形象，直至唐代，柜都是一种长方形大箱带足的样式，柜门较小，设在上方。唐墓多出土三彩陶柜，日本正仓院则藏有不少这种带足的箱式柜，且都记为“柜”。较早的竖柜形象，有正仓院所藏“赤漆文欟木御厨子”，为包括天武天皇在内六代天皇传承之物，相当于我国盛唐时期之物，可见橱柜初时的样式，但是这种柜子在国内暂无实物和资料发现。辽墓壁画和宋画中可见一种前开门的盝顶小柜，可能是国内竖柜的较早期形象。实物而言，暂未发现明代以前对开门竖柜的资料。明清时期的柜品类丰富，有亮格柜、万历柜、圆角柜、方角柜、顶箱柜等。

The cabinet (*Gui*), originally called “*Kui*”, is vividly restored by the pottery miniature of cabinet unearthed from the Tomb of Han Dynasty (206 B.C.-220 A.D.)at Zhang Wan Village in Lingbao County of Henan Province, and the image of a lying cabinet carved in the stone relief of Han Dynasty (206 B.C.-220 A.D.) at Yinan County of Shandong Province. Until the Tang Dynasty (618-907), the cabinet maintained as a large rectangular box with feet and small doors set on the top, like many tri-colour glazed miniature pottery cabinets unearthed from the tombs of Tang Dynasty (618-907) and several chest-styled cabinets (with feet) collected in Shōsō-in (temple) in Japan, which are all recorded as “cabinet”. The image of early vertical cabinet can refer to the “red lacquered wooden imperial cabinet” stored in Shōsō-in (temple), which is a rare existing early cabinet inherited from six emperors; however, none of any similar cabinets has ever been found in China except for the image of a small cabinet with truncated pyramid top and front doors in murals of the tomb of Liao State (907-1125) and paintings of Song Dynasty (960-1279); and the earliest vertical cabinet found in China was made in the Ming Dynasty (1368-1644). In the Ming and Qing Dynasties, the cabinet evolved to many kinds and variations including shelf cabinet, Wanli-styled cabinet, round-cornered cabinet, square-cornered cabinet, and combined cabinet.

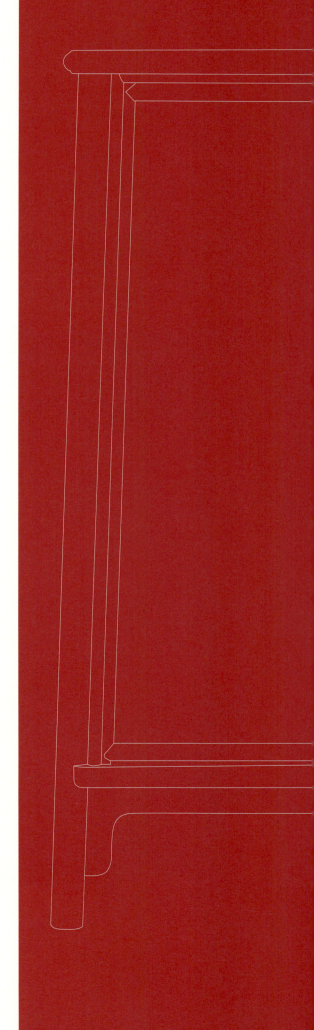

098

有闩杆圆角柜

Round-cornered Cabinet with Latch Post and Pivot Doors

69×43×97cm

此柜为典型的圆角柜造型。柜帽四面攒边，中装薄板，边抹做泥鳅背压双边线线脚处理。柜帽向外喷出一部分，下安四腿，腿内方外圆，四腿八挓，收分明显，形成下大上小的稳定感。

柜帽下安两柜门，柜门中间安闩杆，柜门中间和闩杆上安条形面叶，上安吊牌和钮头，可以把柜门和立柱闩在一起，便于锁牢。柜门四面攒边，两柜门心板用整板对开，纹理对称。两柜门外边向下延伸，形成木轴门，安装在腿间横枨上挖出的臼窝里。

横枨之下安刀板牙子，光素无饰。

此圆角柜造型比例精巧，简洁简素，柜门心板用整板对开，纹理对称优美，成为整件圆角柜的视觉中心。

This is a typical round-cornered cabinet, with top panel braced with assembled frames with edges polished in loach-back style and planed down two rabbets on both sides as borderlines; four splayed legs are installed indentedly below the top panel to form a steady structure.

Two pivot doors are installed below the top panel along the splayed legs; a vertical latch post is butted in the middle between two door panels; rectangular latch bases are inlaid on edges of door panels and the latch post with escutcheons and handles; door panels are constructed with assembled frames and two central boards made out of the same one-piece material, hence the symmetrical veinings; the far side of two doors are fixed by pivots installed in the sockets gouged in the lower stretcher between legs.

Below the lower stretcher installs unadorned apron plates.

This round-cornered cabinet is of delicate scale and simple structure; its two door panels are made out of the same one-piece material, so as to reveal a beautiful symmetrical veinings, becoming the visual attraction of the whole piece.

099

劵口牙板带栏杆亮格柜

Display Cabinet with Balustrades and Apron Plates

88×43×168cm

此亮格柜由花梨木制成，花梨木色泽淡雅，纹理流畅。

亮格柜做圆角柜式，上面柜帽四面攒边，中装薄板，边抹做泥鳅背压双边线线脚处理。柜帽之下安腿，腿截面内方外圆，四腿八挓，收分明显。腿中间部分安横枨，横枨之上为开敞空间，后背装背板，其他三面装壶门劵口牙子和栏杆。劵口牙子光素，边缘起阳线。栏杆为灯笼加十字纹锦地，细致精雅。

横枨之下正中安闩杆，闩杆两侧装门，柜门四面攒边，中打槽装薄板。柜门之下安横枨，横枨上挖臼窝以放门轴。柜门正中安条形面叶，上安吊牌和钮头。柜门之下装刀板牙子。

此亮格柜上部半开敞，下部封闭，形成上虚下实的柜体空间。造型独特，比例协调，简洁中不失细节。

This display cabinet is made of *Huali* wood with elegant colour, smooth and natural veining.

This is a typical round-cornered cabinet, with top panel braced with assembled frames with edges polished in loach-back style and planed down two rabbets on both sides as borderlines; four splayed legs are installed indentedly below the top panel to form a steady structure. A stretcher is butted in the middle between legs to divide the space; above the stretcher is exposed on the front side, sealed on the back side with a back panel, and enclosed by exquisite balustrade-styled fenders decorated with assembled lantern-and-cross-shaped lattice work and arabesque apron plates of *Kun*-gate style (below the top panel) on three sides.

Below the stretcher installs a vertical latch post in the middle between two door panels; door panels are constructed with assembled frames and two central boards; the far side of two doors are fixed by pivots installed in the sockets gouged in the lower stretcher between legs; rectangular latch bases are inlaid on edges of door panels and the latch post with escutcheons and handles; unadorned apron plates are butted below the door panels.

This display cabinet is built in a quite distinctive structure with proper scale and simple yet delicate decorations, leaving its upper part exposed on the front side to form a unique spacial variation.

100

券口牙板带栏杆万历柜

Wanli-styled Cabinet with Apron Plates and Balustrades

100×43×195cm

此万历柜由花梨木制成，色泽淡雅，纹理流畅。

万历柜上部半敞空间，后背装板，三面壶门券口牙子，落在有望柱的栏杆上。壶门券口牙子曲线翻折有力，边缘起阳线，部分阳线向内翻转出卷草。正面两侧设栏杆，两望柱和横枨围合，中打槽装板，上透雕麒麟纹，下有壶门亮脚。正面中间拉横枨界出下部空间，亦做壶门亮脚。

万历柜下部在横枨之下安闩杆，闩杆两侧装柜门，柜门四面攒边。柜门与柜腿以条形合叶连接。两门中间安条形面叶，安吊牌和钮头。柜门打开后安抽屉。

万历柜下承底座，底座单设，束腰式，下接壶门牙子和三弯腿。

此万历柜上半敞，下封闭，形成上虚下实的空间。比例合宜，造型雅致。

This Wanli-styled cabinet is made of *Huali* wood with elegant colour, smooth and natural veining.

It leaves the upper part exposed on the front side, sealed on the back side with a back panel, and enclosed by arabesque apron plates of *Kun*-gate style outlined with embossed borderlines and floral designs (below the top panel) on three sides; two exquisite short balustrade-styled fenders are butted on the front side, carved with openwork of kylin design and *Kun*-gate apron plates below; between the two sides balustrades connects a slim stretcher carved with bas-relief of floral *Kun*-gate pattern to separate the space.

Below the stretcher installs a vertical latch post in the middle between two door panels; door panels are constructed with assembled frames and two central boards; the far side of two doors are fixed by hinges; rectangular latch bases are inlaid on edges of door panels and the latch post with escutcheons and handles; drawers are installed in the lower cabinet.

The base is constructed with girdled waist, *Kun*-gate apron plates, and three-curved legs (cabriole legs).

This Wanli-styled cabinet is built with proper scale and elegant style, leaving its upper part exposed on the front side to form a unique spacial variation.

101

券口牙板带栏杆万历柜

Wanli-styled Cabinet with Apron Plates and Balustrades

113×45×187cm

此万历柜由白酸枝制成，色泽莹润，纹理流畅细腻。

万历柜上部半敞，安后背板，另外三侧装券口牙子，落在栏杆之上。券口牙子正中浅浮雕拐子纹饰，牙子边缘起阳线，两侧向内翻转成拐子纹。栏杆横直，上安细横枨，中间装板，上浅浮雕双螭纹饰。

万历柜下部装柜门，柜门四面攒边，中装板。柜门与腿以合叶连接，柜门中间安条形面叶，装吊牌和钮头。

万历柜下承底座，底座单设，束腰式，束腰下接壶门牙子和三弯腿。

此万历柜上部半敞，下部封闭，形成稳定造型，加之适当的雕琢修饰，整件家具显得稳重秀雅。

This Wanli-styled cabinet is made of white dalbergia cochinchinensis of refined material with smooth veining and elegant gloss.

It leaves the upper part exposed on the front side, sealed on the back side with a back panel, and enclosed by balustrade-styled fenders carved with bas-relief of abstract dragon designs and arabesque apron plates of *Kun*-gate style carved with bas-relief of fretwork designs and outlined with embossed borderlines (below the top panel) on three sides.

Below the stretcher installs two door panels constructed with assembled frames and two central boards; the far side of two doors are fixed by hinges; rectangular latch bases are inlaid on edges of door panels with escutcheons and handles.

The base is constructed with girdled waist, *Kun*-gate apron plates and three-curved legs (cabriole legs).

This Wanli-styled cabinet is elegantly decorated with proper carvings and bas-relief, leaving its upper part exposed on the front side to form a unique spacial variation and structural steadiness.

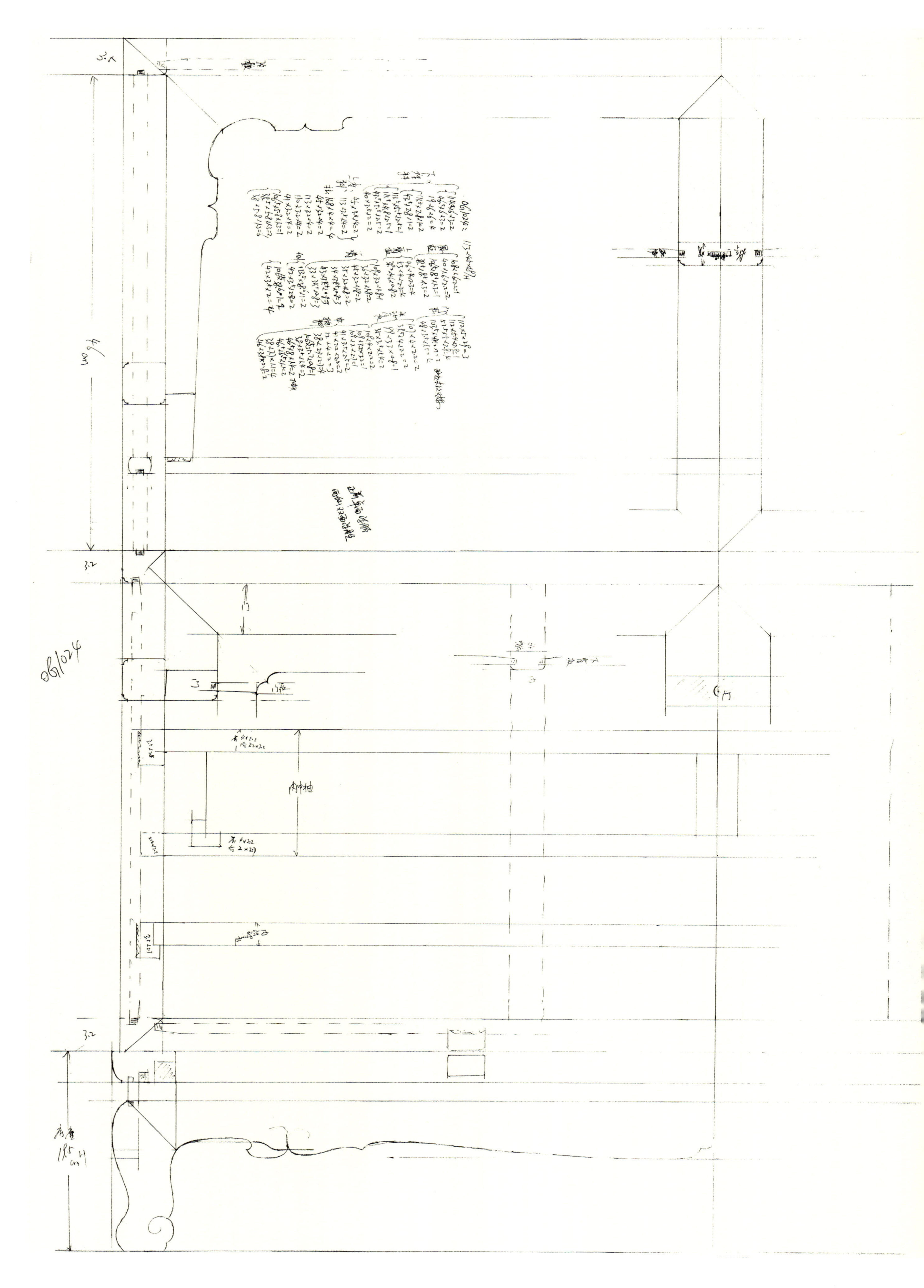

102

螭龙捧寿纹透格门方角柜

Square-Cornered Cabinet with Hinged Doors Decorated with Cross Tracery and Carvings of Two Dragons Holding Up the Chinese Character of "寿" (Longevity)

95×53.5×180cm

方角柜由紫檀制成，紫檀色泽沉穆，纹理流畅，牛毛纹满密。

门六抹，在上中下皆界出窄空间，装绦环板，绦环板上先以阳线为一圈，四角委角处理。阳线之内浅浮雕双螭纹。上部大长方形门板板心上，用攒斗方法做灯笼加十字纹锦地。下部大长方形门板板心上，则浅浮雕螭龙捧寿纹饰。柜门之下安牙子。

柜门与腿以条形委角合叶连接，柜门中间安条形委角面叶，上装吊牌和钮头。

方腿之下包铜包角，既增加坚固，又是合宜的装饰。铜饰件颜色浅淡，与深穆的紫檀形成鲜明对比。

此方角柜造型方正稳定，柜门和侧帮满饰，或攒斗，或浮雕，整件家具富丽堂皇，华贵典雅。

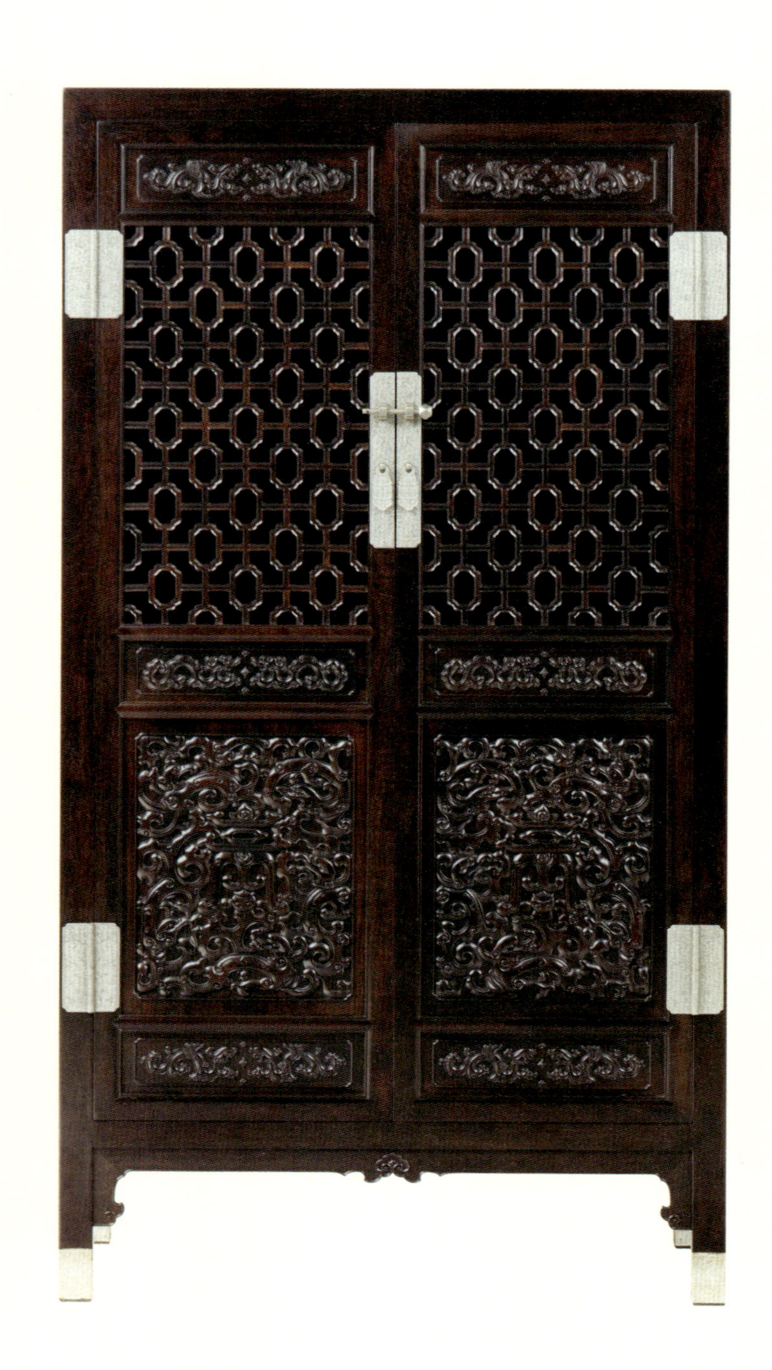

This square-cornered cabinet is made of red sandalwood with dark and solemn colour and subtle ox-hair veining (traces of capillary tubes of the plant).

The two doors is assembled with six door panels to correspond with the cabinet's inner space arrangement, and embedded with decorative waist panels outlined with embossed borderlines and carved with bas-relief of double dragon design framed by a tracery with dented corners; the upper rectangular panels are decorated with assembled lantern-and-cross-shaped lattice work; the lower rectangular panels are carved with bas-relief of two dragons holding up the Chinese character of " 寿 " (longevity); unadorned apron plates are butted below the doors.

Doors are connected to legs with rectangular hinges with dented-corners; rectangular latch bases are inlaid on edges of door panels with escutcheons and handles.

Four feet of square section are wrapped with copper protectors to enhance the steadiness and also to serve as decorations. The light metallic lustre of the copper fittings forms a stark contrast with the dark colour of red sandalwood.

This square-cornered cabinet is built with steady structure, with door panels and side boards fully decorated with assembled latticework or bas-relief, demonstrating the luxurious grandeur and solemnity.

103

顶箱柜

Combined Cabinet

110×60×235cm

顶箱柜以黄花梨制成，选料甚佳，赤红色若火焰，质感温润可人，光素无装饰和方正的造型，将黄花梨的质感之美充分展现。

柜为四面平结体，上为顶箱，下为竖柜，上下柜门心皆以独板对开而成，山峦流水般纹路，两两对称，这是自晚明以来家具审美特征的延续，以自然朴实的木纹之美作为装饰。中设闩杆，装白铜簇云纹合叶，以铜泡钉固定其上，吊牌为鱼形变体。铜饰件造型优美，与黄花梨相互映衬，颇具画龙点睛般的装饰效果。竖柜亦为四面平式，铜饰件尺寸变大，安装位置根据比例重新调配，形成协调自然的布局。柜下设闷仓，内有搁板，下方可放置贵重物品。柜下四面装刀牙板，简洁素气，和顶箱柜整体相合。下设铜套足，可防止腿足材料开裂，耐磨耐蚀，使得顶箱柜更加坚固牢靠。

This combined cabinet is made of yellow *Huali (Huanghuali)* wood of refined material with graceful gloss and flaming red colour; the beautiful texture of yellow *Huali (Huanghuali)* wood is fully brought out by its unadorned design of simplicity.

The cabinet is built in flat-corner structure (corners butted with mitered corner bridle joints, also called cubic-cone rice-pudding joint), with upper hat chests and lower vertical cabinets; all the door panels are made out of the same one-piece material, hence the symmetrical veinings, which is the aesthetic character of late Ming Dynasty (1368-1644): to decorate with the natural veining of the material. A latch post is installed in the middle and inlaid with cupronickel latch base carved with cloud design installed with escutcheons and fish-shaped handles. The copper fittings form a beautiful contrast with yellow *Huali (Huanghuali)* wood. The lower vertical cabinets are also built in flat-corner structure (corners butted with mitered corner bridle joints, also called cubic-cone rice-pudding joint), with larger copper fittings to accommodate their size. A hidden compartment is set at the bottom divided by small panel to store valuable things or items; unadorned apron plates are butted below the cabinet to echo the general simple style; four feet are wrapped with copper protectors to prevent them from cracking, friction and erosion.

104

灵芝纹牙板顶箱柜

Combined Cabinet with Apron Plates Carved with Ganoderma Patterns

115×60×230cm

顶箱柜以黄花梨制成，成器已经年，故而黄花梨光泽更加内蕴，有薄薄一层包浆，火气退尽，与初制的黄花梨艳丽之美不同，有一份稳重内敛之美。

此柜线脚变化，要比前件复杂一些，柜框素混面，即所谓的“泥鳅背”“指甲圆”效果，内侧边缘起打洼皮条线装饰，门框亦是相同线脚。柜门为落膛起鼓式，双拼板，对开而成，花纹对称，山岚纹和大流水纹为主。此柜不再设闩杆，为硬挤门式，整体比例较秀气。

柜下装壸门牙板，牙头处勾卷云纹，沿边起线，至中间壸门处相交后衍为卷草纹，枝蔓上各生出两个灵芝头，点缀活泼，寓意吉祥，壸门下雕饰分心花。

设委角长方形铜面叶、合叶，其上錾刻夔龙纹，并将纹饰外的地子錾为珍珠纹锦地，颇费功夫。吊牌为变体如意式，亦錾刻花纹。

This combined cabinet is made of yellow *Huali (Huanghuali)* wood of refined material with graceful gloss and a thin layer of oxidation film due to its long history; compared with the bright colour of fresh cut yellow *Huali (Huanghuali)* wood, the ancient wood reveals a subtle elegance.

The moulding of this combined cabinet is more complicated than the previous one: edges of the cabinet frames are polished in loach-back style and outlined with embossed belt-lines (wider than the usual borderlines) gouged out a round low-lying groove in the middle along the inner rims of the frames; door frames are also decorated with the same moulding; door panels are assembled with double boards made out of the same one-piece material with symmetrical veinings (mostly including continuous mountain patterns and running water patterns); without the latch post, the cabinet appear more slender and elegant.

Arabesque apron plates of *Kun*-gate style carved with floral bas-relief of ganoderma patterns in the middle and cloud designs at corners indicates auspicious meanings; and a flower crocket is attached below the *Kun*-gate apron plates.

Rectangular copper latch base with dented corners is inlaid along the side of door panels, carved with dragon patterns on pearl ground, which is very complicated and time-consuming, installed with *Ruyi*-shaped handles carved with patterns.

105

云龙纹顶箱柜

Combined Cabinet with Bas-relief of Dragons and Cloud

141×68×282cm

顶箱柜以紫檀制成，色泽沉穆，其上金星浮动，深色的紫檀与明亮的白铜饰件相映衬，清新跳脱。

柜框平素，仅转角微微倒棱。门框内侧起圆润的阳线，将满雕的门心和门框界开。门心深浮雕云龙纹，连绵起伏的云纹密密匝匝，神龙显现其间，张嘴扬鬣，指爪撑开拿云，腿上火焰翼飘动，神采奕奕，仿佛正在将祥瑞撒播于人间。其中顶箱上雕刻相对的两个降龙，竖柜上雕刻一升两降计六龙翻腾于云间，闷仓板上雕刻行龙一条。合集为九龙之数，九为单独数字中最大，且为至阳，又暗含九五之尊的寓意。柜下牙板上亦雕两头相同的行龙，组成双龙戏珠纹，远小于上方的九龙，主要作为装饰效果而存在。柜侧落膛起鼓装板，委角长方形开光内深浮雕云龙纹，与正面呼应。柜上附白铜錾云龙纹合叶、面叶，吊牌上饰云纹。

此顶箱柜外框简素而门心繁华，简繁对比强烈，主体分明，形成繁而不乱的装饰效果，不失礼法、秩序，庄严稳重，是清乾隆家具风格的完美体现。

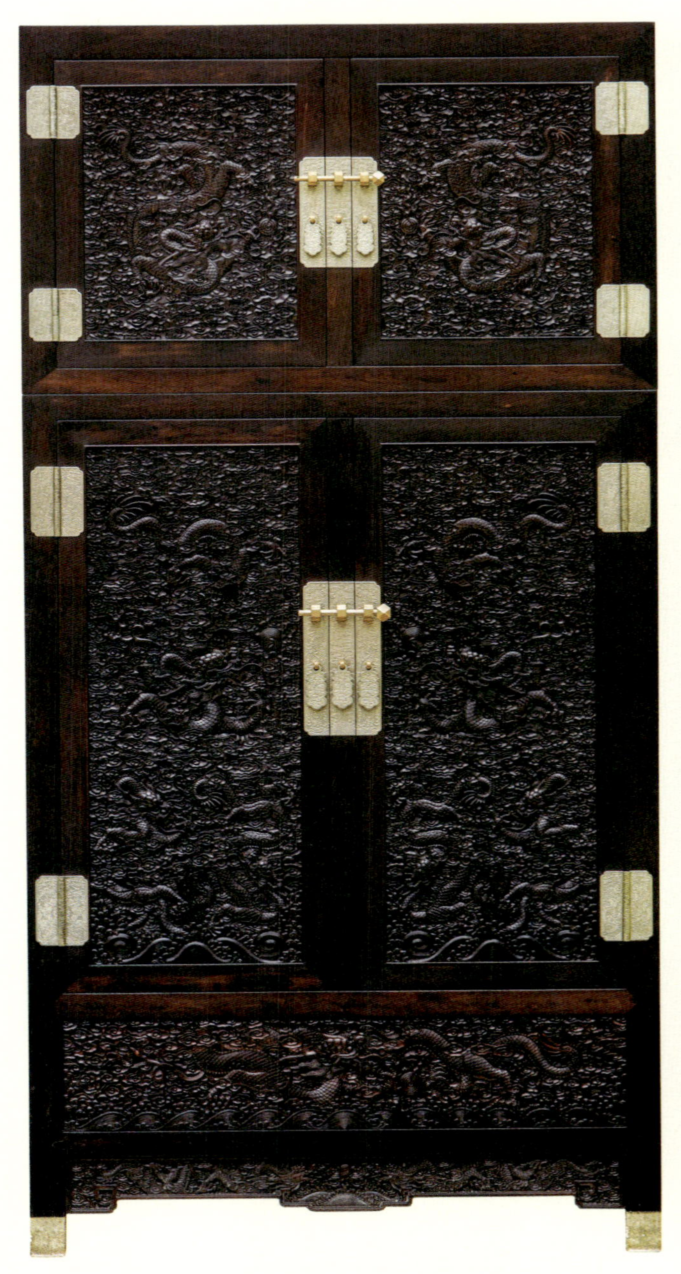

The combined cabinet is made of red sandalwood with dark solemn colour and subtle golden pinhole veining (traces of capillary tubes of the plant), forming a stark contrast with bright cupronickel fittings.

The cabinet frames are unadorned and slightly chamfered at corners; the inner rims are outlined with embossed borderlines to separate the fully carved central door panels and frames; door panels are densely carved with bas-relief of dragon and cloud designs: dragons looming among clouds, with open mouths, stretched claws and flaming fire flickering along, seemingly to spread blessings to the earthy world; two opposite descending dragons are carved on each door panel of the hat chest; each vertical door panel is carved with one ascending dragon and two descending dragons among the cloud; and a horizontal running dragon is carved below on the panel of the hidden compartment; totally nine dragons. Since the number nine is the largest digit and indicates the extreme masculine, nine-dragon design often implies the emperor. Apron plates butted below the cabinet are also carved with two horizontal running dragons, only in smaller size, to serve as decorations. The side panels are carved with bas-relief of dragon and cloud design framed by rectangular tracery with dented corners to echo the front door panels. Cupronickel hinges and latch bases carved with dragon and cloud pattern, and handles decorated with cloud patterns are inlaid on the door panels.

This combined cabinet is built with unadorned frames and luxuriously decorated door panels, forming a striking contrast and bringing out the main body of the furniture, demonstrating a majestic solemnity, as a perfect representation of Qianlong style in the Qing Dynasty (1644-1911).

106

鸾凤牡丹纹顶箱柜

Combined Cabinet with Bas-relief of Paired Phoenixes and Peony Flowers

110×60×220cm

顶箱柜由花梨木制成，选料甚精，纹、色基本一致。

宽为高之半，两柜并列为正方形，厚为宽之半，颇具法度。

外框平素。柜门为落膛镶框做法，边框内侧起有压边线的委角线。柜门为厚板，上深浮雕图案装饰，饰太湖石、牡丹、鸾凤纹，点缀有兰草、菊花，鸾凤互相召唤，嬉戏穿梭于牡丹花下，是传统的“凤穿牡丹”图案。顶箱图案布局与底柜类似，细节又有变化，鸾和凤作盘旋飞舞状，回首相顾。鸾凤周围点缀有连云纹，采用平底起坡的雕刻方式，边缘铲阴线强调，层次分明。柜下有闷仓，闷仓雕双鸾对飞，中饰四角云纹。成对的顶箱柜各有十只鸾凤，合计二十只，概有“十全十美”之寓意。

牙板壶门式，有转折并翻小牙，为清早期常见牙板样式。牙板边缘起阳线，至中间交缠后演化为缠枝莲纹。柜侧山板采用落膛起鼓做法。

This combined cabinet is made of *Huali* wood of highly selected material with consistent veining and colour.

The width is half of the height, so two cabinets can make a square; and the depth is half of the width; which reflect its strict scale and well-balanced proportion.

The cabinet frames are unadorned with the inner rims are outlined with embossed borderlines; door panels are densely carved with bas-relief of literati rock (Taihu lake stone), peonies, and phoenixes flying among peonies, chrysanthemums and other flowers, as typical traditional "phoenix flying among peonies" pattern. The panels of hat chests are carved with similar patterns, yet with small variation: phoenixes circling above and dancing, playing and echoing with each other; the ground is carved with cloud patterns which are emphasized by incised lines, full of depth and spirite. A hidden compartment is set below the cabinet, with its panel carved with double phoenixes and cloud patterns. Each cabinet is carved with ten phoenixes, so a pair of cabinets have total twenty phoenixes, implying the blessing of "being perfect".

Arabesque apron plates of *Kun*-gate style outlined with embossed borderlines are butted at the bottom, representing the common type of apron plates in the early Qing Dynasty.

箱盒

Chest/Case

箱在古代有多种名称，如箧、椟、匮、函、匣、盒等，虽所指大小造型稍有差别，但都是箱的范畴。明清时期的箱，主要有长方箱、官皮箱、书箱、药箱、轿箱等。长方箱是最常见的品种，多设铜合叶、面叶，可以上锁。官皮箱是一种有盝顶或平顶盖，前设屉门或对开门，内有抽屉的箱具，有的上屉中还有镜架，可知是能当做梳妆用具的，故又名"鉴妆""捡妆"等。书箱是中大型带提梁的箱，可搁置案上，收纳书籍等文房用品。药箱是对有抽屉、提梁的小箱的统称，其实不一定是医药用具，也有搁置佛像或文玩者。轿箱则是一种倒"凸"字形的长箱，搁置在轿子前方，是坐轿人随身携带搁置物品的用具。盒则是小型的箱，多为放置贵重物品之用，也有放置名帖的帖盒，放置食物礼品的捧盒和放置书画的画匣。

In ancient times, chest was refered by many names, such as 箧 (*Qie*), 椟 (*Du*), 匮 (*Kui*), 函 (*Han*), 匣 (*Xia*) and 盒 (*He*), etc., which are varied in different sizes. In the Ming and Qing Dynasties, the chest mainly includes rectangular chest, official suitcase, bookcase, medicine casket, sedan case, etc.; and the rectangular chest is most commonly seen, inlaid with copper hinges and latch bases; the official suitcase is with truncated pyramid top or flat top, front doors or drawer doors, even the mirror stand in the upper drawer which can be used as toilet articles; the bookcase is usually built in medium or large size (can be placed on the narrow table), with handle and used to carry books and other stationeries; the medicine casket is the general name for small case with handle and drawers, and is used to carry medicines or sometimes Buddhist statues or antiques; the sedan case is built in reversed "凸" shape, with longer case body and usually placed on the front of the sedan chair for passengers to carry their belongings; the case is refered to small chest, mostly used to store valuable items; moreover, there are invitation case used to carry letters, food case used to contain snacks or food, and painting case used to store calligraphies and paintings.

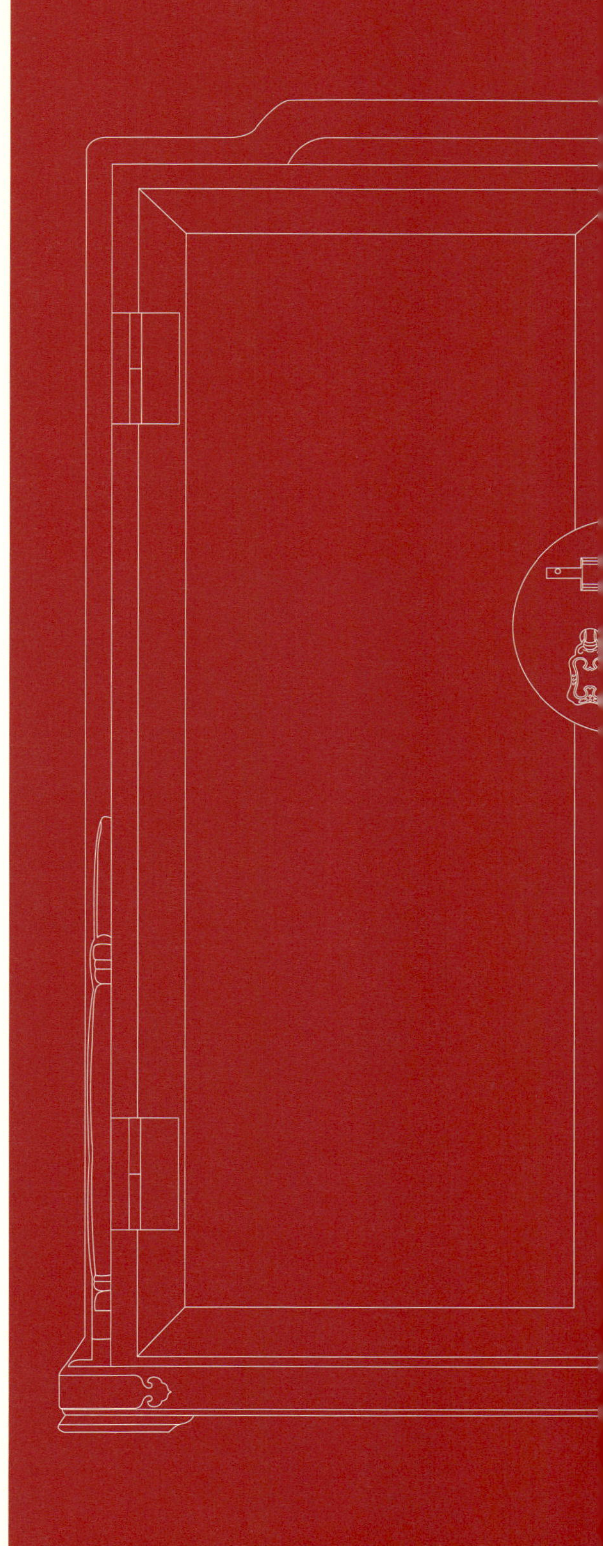

107

云龙纹平顶官皮箱

Flat-top Official Suitcase with Bas-relief of Dragons and Cloud

40×29×35cm

官皮箱紫檀制成，平顶，上盖前脸浅浮雕双螭纹，盖下有平屉。下安两柜门，两扇门对开，门上留子口，顶盖关好后，扣住子口，两门就打不开了。两柜门和上盖上安面叶，面叶上安钮头和云头形拍子，铜活之上亦有缠枝莲纹饰。柜门四面攒边，心板上做壶门开光，开光内外皆浅浮雕云龙纹饰。两柜门与立墙以铜制合叶连接。柜门之间安面叶，上安吊牌。

下为矮台座，台座正面做壶门曲线，上亦浅浮雕双螭纹。

此官皮箱为典型的平顶官皮箱造型，独特之处为正脸的满雕纹饰，浅色的金属饰件和深穆的木材形成鲜明对比。

The official suitcase is made of red sandalwood, with flat top carved with bas-relief of double dragons pattern, flat drawer below the top panel, and a small cabinet with two doors (doors are bored with paired groove to serve as lock) at bottom; two cabinet doors and the top lid are inlaid with latch bases set with escutcheons and cloud-shaped handles; all the copper fittings are decorated with bas-relief of lotus patterns; doors are constructed with assembled frames and central panels carved with bas-relief of dragon and cloud patterns and *Kun*-gate style traceries, and connected to side panels by copper hinges; latch bases are inlaid on the edges of two doors and installed with handles.

The base is processed in *Kun*-gate style, and carved with bas-relief of double dragons.

This is a very typical official suitcase with flat top lid, with unique densely carved bas-relief of dragons and cloud, and light colour metal fittings forming a striking contrast with the dark and solemn red sandalwood.

108

云龙纹平顶官皮箱

Flat-top Official Suitcase with Bas-relief of Dragons and Cloud

40×29×35cm

官皮箱以黄花梨制成，尺度适宜，正面雕刻装饰繁而不乱，是秾华风格家具中的精品。

平顶，箱盖四墙平直，正面陷地浮雕双螭纹，双螭相背而行，回首相望，头如鲶鱼，杏核眼，蒜头鼻，独角，须发四绺，口衔卷草纹，身躯若虎豹，夭矫灵动，四足蹬开，尾巴衍为卷草纹。柜门上浮雕菱花形开光，其内雕升龙，张口扬髭，眼如灯泡，喜气洋洋，身躯如蛇，四足蹬开，周匝点缀如意云纹。菱花形开光外的三角形岔角处，各雕一个螭龙纹，造型与箱顶相近。柜的底座正面挖为壶门牙板，其上浮雕双螭，随着牙板曲线的起伏而相向奔腾，壶门尖处下垂一朵灵芝，两旁壶门线条下，透雕卷草纹。

设白铜面叶、拍子、合叶等，上錾刻缠枝莲纹。其他左、右、背三面一任光素。

此官皮箱正面雕饰纹路最为引人瞩目，繁而不乱，图案风格明风浓郁，质朴自然，是工艺繁杂而艺术水平甚高的佳品。

The official suitcase is made of yellow *Huali (Huanghuali)* wood of refined material, with proper scale and the front side and door panels densely carved with bas-relief of dragons, representing the essence of luxury style in traditional Chinese furniture.

It's constructed with flat top and four vertical side plates; the front panel is carved with bas-relief of double dragons patterns: two dragons with catfish-like heads, almond-shaped eyes, garlic-shaped nose, single horn, four flocks of hairs, tiger/leopard-like body, and floral tails, having their mouths holding a branch of flowers, and their claws stretching out, flying toward opposite direction and turning their heads back to look each other; the door panels are carved with bas-relief of lotus-shaped tracery and an ascending dragon within; two ascending dragons are with bulb-like eyes and snake-like body, having their mouths open and their claws stretching out, with *Ruyi*-shaped cloud decorated around; the spandrel areas outside the lotus-shaped tracery are carved with a small dragon pattern, resembling with the carvings on the flat top. The base of the cabinet is carved with openwork of arabesque apron plates of *Kun*-gate style and decorated with bas-relief of double dragons design; a ganoderma-shaped crocket is attached below the centre of *Kun*-gate apron plate.

Cupronickel hinges and latch bases carved with lotus pattern are inlaid on the door panels, leaving the other three sides unadorned.

The bas-relief carvings on the front side of the official suitcase are quite striking and luxurious, with strong characteristics of Ming Dynasty (1368-1644), representing the highest quality of craftsmanship and aesthetics.

109

盝顶官皮箱

Official Suitcase with Truncated Pyramid Top

24×18×26cm

官皮箱以黄花梨制成，色黄如蜜，质感温婉，通体光素，以自然木纹为装饰。

盝顶，坡度较常见者略小，更觉方正。对开门，攒框平镶木板心，柜门选料甚佳，黄花梨木纹如火焰窜动，点点斑纹，甚为艳丽。底座较高，立板而成，正面挖壸门式亮脚，两头内卷为卷云纹足。打开箱门，则可见大小高低不同的抽屉五具，可收纳各式物品。

官皮箱在古代陈设案头，可放置印章文书珍玩等宝贵物品，也可放置梳妆用具，前设铜拍子、锁鼻，上锁后随身携带，使用方便，有点类似今人用的公文包。

The official suitcase is made of yellow *Huali (Huanghuali)* wood of refined material, with honey-like yellow colour, warm and elegant gloss, and beautiful natural veinings, without any decorations.

The truncated pyramid top, is with mild slope and square shape; the front door panels constructed with assembled frames and central panels which are highly selected yellow *Huali (Huanghuali)* wood with beautiful flaming veining and scattered flecks; the base is relatively high and with its front side carved into arabesque apron plates of Kun-gate style with cloud design at corners. Open the doors, five drawers of various sizes are arranged neatly inside to store small items and things. The official suitcase is commonly placed on the table to contain seals, documents, or antiques, as well as toilet articles. Copper latch bases, escutcheons and handles are installed on the front side for security concern; it's portable and quite convenient, resembling to the present briefcase.

明万历
《列女传》插图
鲍氏知不足斋刊本

Wanli Period of Ming Dynasty (1573-1620)
Illustration of *Biography of Famous Women in Ancient China*, Zhibuzuzhai's edition from Family Bao

110

百宝嵌婴戏图盝顶官皮箱

Official Suitcase with Truncated Pyramid Top and Multi-precious-objects Inlaid Design of Playing Children

34×25×36.5cm

官皮箱由紫檀制成，紫檀色泽沉穆稳重，又搭配百宝嵌和铜饰件，丰富多彩。

官皮箱盝顶，顶盖面四角镶钉云纹铜饰件，立墙四角用铜叶包裹。顶盖之下安两柜门，柜门四面攒边，面心平装薄板。顶盖下部和两柜门上部安圆形面叶，上安钮头和云头拍子。两柜门与立墙以合叶连接。柜门正中间安条形面叶，面叶上安吊牌。柜门前脸和顶盖前脸百宝嵌婴戏图，在碧波粼粼的湖边，栏杆曲护，山石嶙峋，古树遒劲，几婴童戏耍，憨态可掬，生动传神。

官皮箱下承底座，底座方正，前脸开壶门式曲线，上嵌双螭纹。底座立墙四角用铜叶包裹。

此官皮箱为盝顶带底座，前脸以百宝嵌工艺嵌婴戏图，成为此件官皮箱出彩之处。

The official suitcase is made of red sandalwood with dark solemn colour, inlaid with multi-precious stones and decorative objects and installed with copper fittings.

It has a truncated pyramid top with its upper side decorated by cloud-pattern copper mounts at four corners, and its four vertical edges wrapped by copper leaves; below the top lid, there installs two door panels constructed with assembled frames and central boards; latch bases are inlaid on the lower edge of top lid and upper edges of two doors to make a full round shape; a cloud-shaped escutcheon and a latch pin are installed as well; doors are connected to the suitcase by hinges; rectangular latch bases are inlaid on the doors with H-shaped handles; the front side is fully inlaid with multi-precious stones to express a traditional auspicious image of playing children: handrails are twisting along the lake to separate the scene with rugged rocks, ancient trees, and lovely children who are playing games.

The base is square and upright, with its front side carved into arabesque apron plates of *Kun*-gate style inlaid with double dragons pattern with precious objects; four vertical edges are wrapped by copper protectors.

This official suitcase is built with a truncated pyramid top and a base; its front side is fully decorated with the image of playing children by treasure-inlaid craftsmanship.

111

带提梁书箱

Bookcase with A Handle

60.4×36.2×72.2cm

此书箱由红酸枝制成，精心选料，色彩莹润，纹理舒朗大气。

书箱安在底座之上，底座四面攒边，边抹做混面压边线线脚处理。四角皆用铜叶包裹，四角之下收小足。

底座之上为箱体，箱体前两开门。柜门四面攒边，面心平装薄板。心板为一木对开，纹理对称，皆流畅优雅。箱门与侧墙以铜合叶连接。箱门正中安圆面叶、吊牌和钮头。

底座之上、箱体侧墙两侧树立柱，用站牙抵夹，站牙透雕卷草纹饰。上安横梁，做罗锅曲线，以形成横梁下提手空间。

此书箱为提盒式，体量较大，移动时应为杠抬方式。

This bookcase is made of red dalbergia cochinchinensis of selected material with warm gloss and natural veinings.

The bookcase is attached to a base constructed with assembled frames with edges processed into cambered surface with two rabbets along top and bottom rims; four corners are wrapped with copper leaves and installed with small foot pads.

Its main body is attached above the base, with two door panels installed on the front side, which are constructed with assembled frames and central boards; the central boards are made out of the same one-piece material, hence the symmetrical veinings; door panels are connected to the case by hinges; half round latch bases are inlaid on the front doors to make a full round shape and also installed with handles and a latch pin.

Two posts are installed next to the side panels of the case above the base, and clamped by standing plates carved with floral openwork; above the top panel attaches a hunched rail as the handle to leave a space in between.

This bookcase is built in handled tiered case, in large size, assumedly needs to be carried on shoulders.

112

百宝嵌博古图箱

Chest with Multi-precious-objects Inlaid Design of Antiquity Appreciation

38×20×38cm

宝箱由紫檀制成，紫檀色泽沉穆稳重，结合百宝嵌工艺，在色彩上形成鲜明对比，在质感上也有很明显区别。

宝箱为异形，上带箱帽，箱帽四面攒边，中镶板，边抹做泥鳅背压边线线脚处理。箱帽之下内收以短柱围合出抽屉空间，抽屉脸以百宝嵌做折枝花果纹饰，正中安圆形铜把手。

抽屉之下安似束腰下的托腮，再置箱体，是抽屉向箱体的过渡。箱体正面对开门，箱门四面攒边，面心装板，板上以百宝嵌做博古图纹饰。

箱体之下再向内收，安与箱体之上相似的抽屉。下抽屉之下安底座。底座座面四面攒边，下收束腰，束腰之下安三弯腿和牙子。

此宝箱上下两抽屉，中间为箱体，下承底座，造型奇特，非典型制式。

The chest is made of red sandalwood with dark solemn colour, inlaid with multi-precious stones and decorative objects to form a striking contrast in colours and textures.

It has a special-shaped top panel (chest hat) which is constructed with a central board and assembled frames with edges polished in loach-back style and planed down two rabbets on both sides as borderlines; below the chest hat installs a drawer enclosed by short struts, with its front side inlaid with multi-precious-stones to form images of twigs of flowers and fruits and installed a copper handle in the middle.

Below the drawer installs a girdled waist and the chest chamber with two door panels constructed with assembled frames and central panels which are inlaid with multi-precious stones and objects to present a design of antiquity appreciation.

Below the chest chamber installs the same drawer structure as the upper one, and the base which is braced by assembled frames and attached to a girdled waist below; four three-curved legs (cabriole legs) and apron plates are butted into the bottom.

This chest is built in a rare and quite distinct structure, with a pedestal base and two drawers above and below the chest chamber.

113

百宝嵌五福捧寿纹箱

Chest with Multi-precious-objects Inlaid Design of Five Bats Clustering Round a Chinese Character "寿" (Longevity)

35×22×32cm

此宝箱由紫檀制成，表面以百宝嵌做不同纹饰，在沉穆紫檀之上显得分外耀眼。

宝箱整体造型做方形拐角，凹凸变化丰富。前脸上面以硬拐子和短柱围合出两抽屉的空间，抽屉脸嵌出双蝠纹，正中安拉手。之下为工字形箱体，中安闩杆，闩杆两侧安凹形箱门，门体鼓出两块安合叶与侧墙相接。箱体攒边，内装板，上嵌五蝠捧寿纹饰。

箱体之下安两抽屉，抽屉分置两侧，中空。抽屉脸嵌双螭纹，正中安拉手。抽屉之下安小足。

宝箱侧面和上面接嵌出拐子纹、螭龙纹、五蝠纹等纹饰。

此宝箱异形，共设三抽屉和两柜门，表面以百宝嵌做各种纹饰，在紫檀深色表面之上形成丰富的色彩变化和质感变化，整件宝箱显得华贵富丽。

The chest is made of red sandalwood and inlaid with multi-precious stones and decorative objects to form a striking contrast with the dark solemn colour of the material.

The chest is built in a quite unique structure with straight hard turns and concave-convex variations. The front side of the upper drawer is divided by fretwork and short struts to indicate its position, and inlaid with multi-precious stones to form the image of double-bat design, with two handles in the middle. Below the drawer is the H-shaped chest chamber with a latch post installed in the middle between two door panels; doors are connected to the side panels by hinges; and the chamber is constructed with assembled frames and central panels inlaid with design of five bats clustering round a chinese character " 寿 " (longevity).

Below the chest chamber installs two drawers on each side; the front side of the lower drawers is inlaid with double dragons pattern and installed a handle in the middle; four small foot pads are attached to each lower drawer.

The side and top panels are inlaid with abstract dragon motif, double-dragon pattern, and five-bat design.

This chest is built in a rare and quite distinct structure, with three drawers and two cabinet chambers; the front side, the top and side panels are fully decorated with multi-precious-objects inlaid patterns and designs, creating various changes in colour and texture on the dark ground of red sandalwood to make the whole chest an absolute opulent luxury.

114

百宝嵌八宝纹箱

Chest with Multi-precious-objects
Inlaid Design of Buddhist Eight Treasures

35×22×32cm

此宝箱由紫檀制成，在深穆的紫檀表面，使用百宝嵌工艺，嵌出佛八宝、五蝠纹、双螭纹等各种纹饰。

宝箱异形，以方形为基础，两侧凹凸。前脸上两具抽屉并置，下两具抽屉分列，中间为对开门。门和抽屉周圈皆做线材硬拐子，素混面线脚处理。上抽屉脸嵌出双蝠纹，两柜门上嵌出佛八宝，八宝分置两门，上下再嵌祥云以做装饰。下抽屉脸嵌双螭纹。箱体侧面嵌出拐子纹饰、五蝠纹、双螭纹等纹饰，为骨嵌，颜色乳白，与深穆的紫檀形成鲜明对比。

此宝箱造型特意，异于常规，紫檀表面以百宝嵌满嵌各种纹饰，显得富丽堂皇。

The chest is made of red sandalwood and inlaid with multi-precious stones and decorative objects to form images of Buddhist eight treasures, five-bat and double-dragon patterns against the dark solemn colour of the material.

The chest is built in a quite unique structure with straight hard turns and concave-convex variations. Two drawers are juxtaposed above the chest chamber and another two are attached below separately; two door panels are connected to the chest by hinges; the front side of the chest is divided by fretwork and short struts to indicate its inner structure; the front side of the upper two drawers is inlaid with multi-precious stones to form the image of double-bat design; two door panels below are inlaid with design of Buddhist eight treasures and auspicious cloud in the background; the front side of the lower drawers is inlaid with double-dragon pattern and installed a handle in the middle; the side and top panels are inlaid with carvings of bones to form abstract dragon motif, double-dragon pattern, and five-bat design, creating a striking contrast against the dark red sandalwood.

This chest is built in a rare and quite distinct structure, and fully decorated with multi-precious-objects inlaid patterns and designs to make it an absolute opulent luxury.

115

嵌瘿木螭龙纹轿箱

Sedan Case with Burl Inlaid Design of Dragon Pattern

75×17×14cm

轿箱为紫檀嵌瘿木而成，在色彩、肌理上都形成鲜明的对比。

轿箱是在轿子上摆放的箱，两端箱底各缺掉一方块，是为了架搭在亮格轿杠之上。主人坐于轿上，轿箱架于面前。

此轿箱前脸镶瘿木做双螭纹，双螭相对，面向圆形面叶，似在捧圆寿。周边小螭龙或紧随其后，或迎面而上，活泼灵动。

轿箱四角皆以铜叶包裹，既加固，又是适当的装饰。前脸正中的圆形面叶，安云头拍子和钮头。箱盖面上四角镶钉云纹铜饰件。

此轿箱在标准制式基础之上，做瘿木镶嵌出螭龙纹饰，成为此件轿箱的最大特色。

The sedan case is made of red sandalwood and embedded with burl, creating a striking contrast in colour and texture against the dark red sandalwood.

It is a small-sized storing furniture placed in ancient sedan chair, and is built in reversed "凸"shape to fit hanging on the sedan lever in front of the passenger.

The front side of the case is inlaid with burl carvings to form the double-dragon pattern of two dragons surrounded by many small dragons of various postures facing each other and raising their heads toward the round latch bases in the middle as if they are holding up a round "寿" (Chinese character meaning longevity) character.

Four corners are wrapped by copper leaves to reinforce the structure and also to serve as decorations; the round latch bases inlaid in the middle of the front side are installed with cloud-shaped escutcheon; four corners of the top panel are inlaid with cloud-shaped copper fittings.

This sedan case is built in standard structure, and inlaid with burl carvings of abstract dragons to create a unique visual attractiveness.

116

嵌黄杨木双螭捧寿纹提盒

Boxwood Food Case with a Handle and Inlaid Design of Double Dragons Holding Up a "寿" (Chinese Character Meaning Longevity) Character

35.5×20×24cm

提盒由黄花梨制成，镶嵌黄杨木为纹饰。用长方框做底座，两侧树立柱，用站牙抵夹，上安横梁，做罗锅曲线，以增加横梁下提手空间。底座四角用铜叶包裹。

提盒两撞三层，下层盒底落在底座槽内。盒与盖四边皆用铜叶包裹以加固。如此，在两撞三层四角形成贯通的铜叶。盖面四角镶钉云纹铜饰件。

盒前脸和盖前脸镶黄杨做双螭捧寿纹饰，花纹连续，不受分层影响。

盒盖两侧立墙正中打眼，立柱与此眼相对的部位也打眼，用铜条贯穿，以便把盒盖固定在两根立柱之间，铜条一端有孔，还可加锁。

此提盒前脸镶黄杨做双螭捧寿纹饰，为独特之处。

This food case is made of yellow *Huali (Huanghuali)* wood and embedded with boxwood carvings of abstract dragons as decorations; the rectangular base is constructed with assembled frames; two posts are installed next to the side panels of the case, and clamped by standing plates carved with openwork of abstract dragon; above the top panel attaches a hunched rail as the handle to leave a space in between; four corners of the base are wrapped with copper leaves.

The food case is built in three-tier structure; the bottom panel of the third tier is embedded in the base; all the corners and vertical rims are wrapped with copper leaves to enhance the structure; four corners of the top panel are inlaid with cloud-shaped copper fittings.

The front side of the case is inlaid with a continuous design of double dragons holding up a round "寿" (Chinese character meaning longevity) character.

The side panels and vertical posts are drilled with small holes for the long and slim copper latch pin to fix the lid between the two posts; the latch pin is also drilled with a small hole for the lock.

This food case is inlaid with boxwood carvings of double dragons holding up a "寿" (Chinese character meaning longevity) character to create a unique visual attractiveness.

117

盝顶长方盒

Rectangular Case with Truncated Pyramid Top

29×17×8.6cm

盒由紫檀制成，紫檀色泽沉穆，纹理细腻，牛毛纹满密。

盒由盒盖和盒身两部分组成。盒盖盝顶式，盖顶四面向上起弧形坡。盒盖侧壁自边缘向上挖出弧形曲线，似罗锅枨，边缘起宽线，以强调轮廓。

盒身边缘亦做罗锅圈的台座，台座略有厚度，以承接盒盖。台座之内安横竖板，围合出长方形的空腔，腔体之内以横竖板界出大小、长短不同的空间，以盛放不同大小的饰品。这些横竖板边缘做双混面线脚处理，增加丰富的细节。

盒盖和盒身扣在一起，盒盖上的罗锅曲线外壁，和盒身上的罗锅曲线外壁将重合成一平面，形成一个长方的、完整的盒子，只留出罗锅曲线突出的阳线轮廓。

此盒造型独特之处在于盒身和盒盖外壁的罗锅轮廓。

This case is made of red sandalwood with dark and solemn colour and subtle ox-hair veining (traces of capillary tubes of the plant).

It is built with a lid and the case body; the lid has a truncated pyramid top with mild slope and is outlined with embossed borderlines to emphasize the hunched-rail-shaped silhouette.

The case body is braced by thick hunched-rail-shaped frames to fit into the lid; plates are installed in the case body to enclose a rectangular chamber in which small plates are inserted to divide the space into different shapes and sizes for various items or accessories; all these plates are processed with double cambered edges to add more details to the case.

The lid and the body are fastened together so to make the hunched-rail-shaped frames fit into each other and form a complete rectangular case, only leaving the silhouette of embossed borderlines.

This case is quite distinct with its hunched-rail-shaped silhouette of the body and lid.

118

云龙纹画盒

Painting Case with Bas-relief of Dragons and Cloud

65×9.5×7cm

长方画盒紫檀制成，紫檀色泽内敛，质感富贵，木质细腻，不易变形，适宜雕刻，可以制作甚为紧密的构件，故而高档的盒、匣乃至书封，都以紫檀木制成。

画盒造型狭长，立板为四墙，短的一侧堵头稍短，上方容纳盒盖，盒盖做成榫舌，通过长向立墙上的槽口抽拉。盖面之上浮雕一周连绵不断的云纹，中间海水翻腾，拥簇着嵌牙题签，上以牙嵌成“珠玉潜水，澜表方圆”八个字。珠玉潜水，比喻君子内秀而不张扬，但是胸中才华还是会慢慢显露，兴起方圆波澜，语出《文心雕龙》。题签上方雕垂蝠纹、泉币纹，寓意福禄。下方海水翻腾中云龙三现，寓意龙行天下。雕刻繁密细致，与雕漆手法如出一辙。

The rectangular painting case is made of red sandalwood with dark and solemn colour and refined texture, giving a sense of opulent luxury; red sandalwood is of high tenacity (not easy to deform), therefore an ideal material for carvings and sculpting; they can be processed into compact components, so that upscale cases, caskets or even letter cassettes are usually made of red sandalwood.

This painting case is built in long and slim structure with straight panels; the short edge is left open on one side for fitting in the sliding cover (tongue and groove joint); the cover is carved with bas-relief of continuous cloud pattern surrounding the water-wave design in the middle; the inscription is carved in the central area inlaid with ivory, as "the emerged treasure pearl, causing ocean water to roil", a metaphor to compare the true gentry to treasure pearl, implying that the humble one with extreme talent would eventually be found and able to exert great influence (from the book *Wenxin Diaolong* written by Liu Xie, in the Southern and Northern Dynasties); above the inscription is carved a hanging bat pattern and water-coin patterns as a blessing for happieness and wealth; below the inscription is carved a dragon looming from the water, implying to travel the world like a flying dragon. The refined and delicate multi-layered bas-relief is densely carved with similar craftsmanship of lacquer engraving.

紫萼垂〻亦
自佳掣風顯
霧霞庭階〻
教依附高枝
上作者應緑
別寄懷

春花試問孰
宜先冒雪寒
梅專信傳高
士美人擬猶
失祇應呼是
地行仙

屏具

Screen

屏具是我国家具门类中甚为特殊的一类，历史悠久，脉络清楚，独成体系。屏具原因礼法而设，是文明发展到一定阶段的产物，并非日常生活所必须，故在家具中它的出现比席、床、案等略晚，但至迟周代已有。屏具的装饰、隔断、遮蔽等功能，是随着历史发展逐渐附加的。最早的屏，有黼依和皇邸。又作依、扆、斧依、黼扆、斧扆等。《礼记·曲礼》载周天子“当依而立”。明清时期，屏具已经极大丰富，主要有座屏和围屏两类。座屏按照使用场所，有床屏、枕屏、桌屏、砚屏、灯屏、地屏等。围屏多为 12 扇，连以铰链，可自由折叠，小者搁置案头，大者 3 米余，屏心多另装书画或漆板，有绦环板、裙板，有黄花梨制者，透雕螭龙纹等图案，玲珑剔透，极富装饰效果。

The screen is categorized as a very special furniture, with a fairly long history, clear lineage and independent system. Originally, due to the development of culture, it appeared as an ornament to represent the strict ritual regulations and social hierarchy instead of a daily necessity, therefore, the screen didn't appear until the Western and Eastern Zhou Dynasty (1046 B.C.-256 B.C.), later than the mat, bed, and narrow table. With the changes of the times, the screen gradually serves more practical functions of decoration, space partition, and shielding. The early screen, called "*Fuyi*"and "*Huangdi*", is recorded in *The Book of Rites*, as "the King of Zhou shall stand next to a screen". In the Ming and Qing Dynasties, the screen has become varied in kinds and sizes, mainly in two categories: pedestal screen and folding screen; in terms of the occasions, the pedestal screen includes bed screen, pillow screen, table screen, ink-stone screen, lantern screen, standing screen, etc.; and the folding screen, mostly with twelve panels hinged together, can be folded at will; the small folding screen can be placed on the table; and the large three-meter-high screen are usually embedded with paintings or lacquer plaques, as well as decorative freize plaques and apron plates; the one made of yellow *Huali (Huanghuali)* wood is commonly decorated with exquisitely carved openwork of dragon patterns.

座屏

Pedestal Screen

顾名思义，座屏即有座的屏，也是屏的最初形态，马王堆一号墓、二号墓皆出土有汉代彩绘座屏模型，为方片式屏扇，下附两个鱼尾式座。其后千余年来，座屏的基本形态基本未曾有大的改变，即使清晚期的座屏，简练者亦基本如此。座屏在席地而坐时代，或设在使用者背后，为身份的象征；或设在侧面，名为“隔座”，起阻挡的作用。明清时期的座屏，多由屏心、屏框、站牙、绦环板、披水牙板、屏墩组成。若是屏扇各异拔插的插屏，则屏座上又增设立柱，上开槽卡住屏扇的竖框。座屏屏心常见有石、木、书画、漆、铜、瓷等质，几乎一切平面装饰都可镶在屏上，其中石最为典型，又见有大理石、祁阳石、玛瑙石等，取法自然，为室内平添几分逸气。

As the name implies, pedestal screen is the screen with a pedestal; it is the most primitive form of the screen, like the colour-painted miniature screens unearthed from the 1st and 2nd Mawangdui Tombs of Han Dynasty (206 B.C.-220 A.D.), which are square screens with two fish-tail-styled pedestal attached below. Since then, after a thousand years, the basic structure of pedestal screen didn't change a bit; even in the late Qing Dynasty, the common pedestal screen are basically built in the same form. In the times when people sat on mats on the ground, the pedestal screen is either placed behind the master to indicate his/her hierarchy and identity, or placed on the side as "seat divider"to shield or separate the space. In the Ming and Qing Dynasties, the pedestal screen is commonly constructed with central plaque, frames, vertical plates, frieze plaques, apron plates, and pedestal; as for the table screen with a removable central plaque, then grooved posts will be installed on the pedestal to fix the plaque; the common central plaque of pedestal screen is usually made of stone, wood, paper or textile with painting and calligraphy, lacquer, copper and porcelain, almost all the two-dimensional decorations; among them, the stone plaque is the most typical, including marble, Qiyang stone and agate, which can bring in the beauty of nature in the living space.

119

镶大理石螭龙纹大插屏

Giant Marble Embedded Screen Carved with Openwork of Dragon Patterns

190×104×219cm

插屏由白酸枝制成，色泽淡雅，纹理细腻流畅。

插屏底座由两块厚木作墩子，下起壶门曲线，表面浮雕草龙纹。底座之上树立柱，以站牙抵夹，站牙上透雕草龙纹。

两立柱之间，安枨子三根，配合三短柱，界出上层三段下层两段，各段打槽装板，透雕草龙纹或双龙捧寿纹。横枨下安八字形的披水牙子，牙子做壶门曲线，上浅浮雕双螭纹。

插屏四面攒边，中用不同长短的横竖材界出大小不同的屏心，上下左右皆留出三块屏心，上装板，透雕草龙纹。正中镶大理石板，其上云气氤氲，纹理入画。

此屏风比例合宜，结构坚固稳定，采用浅浮雕、透雕、圆雕的方式装饰屏风细节，中镶大理石板，气质文雅秀丽。

明晚期
《闵齐伋绘刻西厢记》插图

Ming Dynasty (1368-1644)
Illustration of *Romance of the Western Chamber*, illustrated by Min Qiji

The table screen is made of white dalbergia cochinchinensis with elegant colour and refined veinings.

The pedestal is constructed with two thick panels decorated with bottom *Kun*-gate apron plates and bas-relief of floral dragons; two posts installed on the base are braced by vertical plates carved with openwork of floral dragons.

Between two posts installs three grooved stretchers vertically divided by three short struts and embedded with openwork plaques of floral dragons or the design of double dragons holding up a round "寿" (Chinese character meaning longevity) character; below the bottom stretcher installs a *Kun*-gate apron plate carved with bas-relief of double dragons.

The table screen is constructed with assembled frames which divide several spaces of different sizes for small decorative plaques carved with openwork of floral dragons, and a central marble slab with elegant veinings resembling to ethereal mountains looming in clouds.

This table screen is built in proper scale, with steady structure and so many details decorated by bas-relief, openwork, and circular engravure, and embedded with a marble slab.

明　杜堇《玩古图》
台北故宫博物院藏

Ming Dynasty (1368-1644)
Antique Appreciation, by Du Jin
Taipei Palace Museum

120

镶大理石云龙纹插屏

Giant Marble Embedded Table Screen Carved with Bas-relief of Cloud and Dragons

25×14×32cm

插屏紫檀制成，色泽沉穆，气质内敛，镶嵌大理石屏心，黑白对比强烈，富于装饰效果。

屏扇为委角长方形，里、外起阳线为边，看面浮雕勾云纹，连绵不断，为模仿自青铜器的纹饰。框内又设复框，上浮雕云龙纹，雕刻繁密而工整，复框内为圆光，镶嵌山水纹大理石，宛如明月悬空，清净圆照，暗含哲理。

屏座设立柱，变体方如意头式，卡住屏扇。变体石榴头式站牙，轮廓边缘为夔龙纹，内亦浮雕勾云纹。屏墩为几座式，腿足弯曲，下方挖壸门亮脚，看面亦浮雕勾云纹，中部组合为兽面纹。屏座上下设双枨，夹镶绦环板，内浮雕变体螭龙纹。下方设披水牙板，中间为兽面纹。整体纹饰皆有很强的金石意趣。

此屏为仿制乾隆时宫廷插屏而成，富丽堂皇，尽得皇家气派，中间镶嵌的大理石则在富贵中平添了几分清雅。

The pedestal screen is made of red sandalwood with dark and solemn colour and embedded with marble plaque, forming a striking contrast in colour and texture.

The screen panel is processed in square shape with dented corners, and outlined with embossed borderlines; the edge of front side is densely carved with bas-relief of cloud pattern to imitate the motif on bronze ware; the inner framed area is carved with exquisitely complicated bas-relief of cloud and dragon patterns; in the middle, an round opening is left to reveal the embedded marble slab with beautiful natural veinings resembling to the ink and wash landscape painting; and the round pale white marble slab seems like a full moon hanging in the dark night sky.

The pedestal is installed with two *Ruyi*-shaped posts to fix the screen panel; the pomegranate-shaped vertical plates are outlined with dragon pattern and carved with bas-relief of cloud pattern in the middle areas; the pedestal is built in the form of trestle table, with curved legs, a *Kun*-gate apron plate and front side carved with bas-relief of cloud and beast-face patterns; between two posts installs double stretchers and a frieze plaque in the middle carved with bas-relief of abstract dragon pattern; below the bottom stretcher, installs apron plates decorated with beast-face patterns. All the decorations are derived from the motif from bronze ware.

This screen is an imitation of royal pedestal screen made in Qianlong Period, with opulent luxury and grand solemnity. The central embedded marble slab adds a hint of poetic elegance.

121

百宝嵌博古图插屏

Table Screen with Multi-precious-objects Inlaid Design of Antiquity

107×39×89.5cm

插屏以紫檀制成，镶嵌蜜蜡、青玉、虬角、绿松石等为百宝嵌图案。

屏扇子母框式，外框中间起两炷香线，边起阳线，内框细窄，稍退后安装，密雕回纹。镶嵌紫檀木屏心，其上百宝嵌为博古图，中间宝瓶内插梅花、南天竹，旁置宝石点缀的如意，拴回头穗，寓意岁岁平安如意。周匝又有供盘，内陈设石榴、桃、佛手三果，又称为“三多纹”，寓意多子、多寿、多福。另有莲花、海棠等花卉，皆含吉祥寓意。图案颜色丰富，在深色紫檀映衬下珠光宝气。屏背阴刻折枝杂花纹、点缀蝠纹于其间，亦含良好寓意。

屏座镶嵌绦环板，委角长方形开光内浮雕变体夔龙纹，中间组为兽面纹，设披水牙板，亦雕刻夔龙纹。站牙为宝瓶式，亦浮雕图案，下承以四瓶状足，寓意平平安安。

此屏色泽艳丽，造型比例适宜，雕刻繁而不乱，秩序分明，是模仿宫廷家具的上乘之作。

The table screen is made of red sandalwood, and embedded with beeswax, nephrite, green-tinted walrus ivory, turquoise, etc., forming an image of antique appreciation.

The screen panel is built in the double-frame style: the outer frame is carved with two extra embossed lines (incense lines) in the middle and outlined with borderlines; and the inner frame is outlined with thinner borderlines and densely carved with fretwork; the embedded red sandalwood central plaque is inlaid with multi-precious objects and stones to form the image of antique appreciation: the vase in the middle contains plum blossoms and nandins, with stone-inlaid *Ruyi* (ganoderma-shaped object) lying next to it, tied with auspicious traditional knot to imply the blessing of happieness and peace in each year; around the vase, three offering plates contain pomegranates, peaches and citrus chirocarpus, as "three-blessing-fruit"to respectively indicate fertility, longevity, and happiness; furthermore, there are vases containing lotus flowers, Chinese crabapples, and other flowers to imply auspicious meanings. These inlaid images are of bright and rich colours against the dark red sandalwood ground. The back of the screen panel is carved with intaglio of branches of flowers and bats to express blessings and good wishes.

The pedestal is embedded with frieze plaque carved with bas-relief of abstract dragon framed by rectangular tracery with dented corners; the motifs in the middle area form an image of beast-face; the apron plate is installed below, carved with bas-relief of dragon pattern; vertical plates on two sides are processed in vase-shaped style, and carved with bas-relief; four vase-shaped foot pads are attached to the bottom to imply the meaning of peace and safe.

This screen is inlaid with bright and dazzling colored stones and objects, built in proper scale, and orderly carved with delicate and neat images, absolutely a great imitation of royal pedestal screen.

122

百宝嵌花卉纹座屏

Pedestal Screen with Multi-precious-objects Inlaid Design of Flowers

360×230cm

此座屏风由紫檀制成，紫檀色泽沉穆，配合百宝嵌工艺，在材质上形成鲜明对比。

座屏九扇，下承底座，上覆屏风帽，底座两侧上安两站牙，屏风帽两侧安两挂牙，共同抵夹九扇屏风。

每扇屏风四抹，上下窄段装绦环板，绦环板表面浅浮雕缠枝花纹饰。正中为屏心，上以百宝嵌工艺做四季花卉纹饰，并配御制诗一首。正中一扇最宽，最外侧一扇以横竖材界出左侧三段，只余右侧一段做百宝嵌。

此屏风稳重大气，富丽堂皇，华贵秀美。

原件现展出于故宫博物院家具馆。

The pedestal screen is made of red sandalwood with dark and solemn colour and inlaid with multi-precious stones and objects, forming a striking contrast in colour and texture.

This pedestal screen is constructed with nine panels, a base and nine screen hat plates; the base is installed with two vertical plates on two sides and screen hat plates at two ends are attached with hanging plates to fix the nine panels in between.

Each screen panel is constructed with four parts: upper and lower frieze plaques are carved with bas-relief of floral pattern; the central plaque is inlaid with multi-precious stones and objects to form images of seanonal flowers companied by an inscription of poem written by the emperor; the middle screen panel is the widest in size; the two panels on the far ends are partially embedded with three separate plaques and partially with full-length panel inlaid with multi precious stones.

This pedestal screen is of grand solemnity and magnificence.

Please refer to the screen exhibited in the Palace Museum (Beijing).

架与文玩

Holder/Stand and Antique Objects

相比于前面那些坐卧必备的坐具、承具、卧具等而言，架与文玩更像是“长物”，似非必须，但是营造室内环境，点缀情趣，却非它们不可。架在古代的使用甚为广泛，主要有衣架、灯架、笔架、镜台等。文玩的范围则甚为广泛，似乎只要是文人案头陈设的器具，非吃喝拉撒这些“俗事”所需的物品，都可纳入到文玩的范畴，笔墨纸砚，无所不是。当然本书的讨论范围主要是木器，则收录了香筒、箸瓶、宫灯、案上几及各式圆雕摆件，更值得介绍的是还有两足小型家具模型，皆是将明式家具微缩至三五寸大小的小模型，麻雀虽小五脏俱全，甚耐观赏品鉴，拿在手中，仔细琢磨，又可一窥明式家具造型和结构。

Compared with those daily necessities including seating furniture, storing furniture, and resting furniture, etc., racks and antique objects are more like “superfluous things”instead of requisites; however, regarding to decorations and interior designs, they are of the most significant value. In ancient times, the rack is widely used and mainly includes clothes rack, lantern rack, brush rack and mirror rack (stand), etc.; and antique objects involve even broader range and seems to include all the objects displayed on literati's table (except for those daily necessities concerning people's secular lives), such as writing brushes, ink sticks, paper and inkstones, as well as woodwares dicussed in the following chapter including the incense canister, the chopstick vase, the palace lantern, and various table displays of sculptures and carvings; the most interesting items are furniture miniatures which include all the details and can be held in the palm, without doubt a direct demonstration of the structure of Ming-styled furniture.

架

Holder/Stand

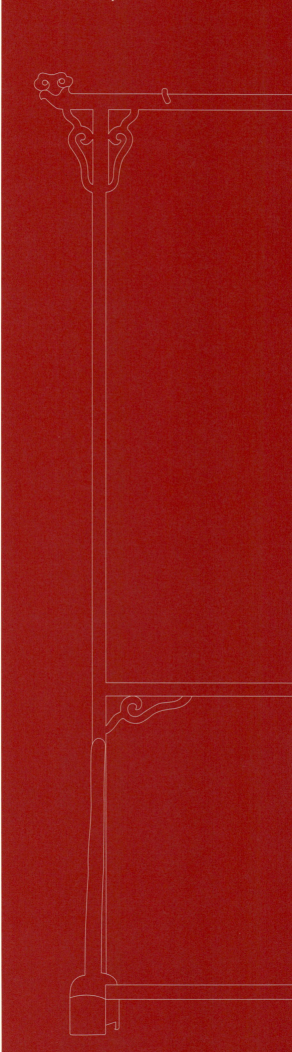

架类收录有衣架、灯架、笔架、镜台和小多宝格。衣架是古人搭扶衣服所用，其造型多妍美，陈设室内，是十分好的装饰。灯架为独梃式，造型亭亭玉立，在古人为必备的用具，当然贫富不同，灯架的精致程度也自不相同。笔架则如同是一件衣架的微缩模型，搁置案头挂笔，是书法练习者不可或缺的用具。至于镜台，在古代应用甚广，其造型为下方若柜体，多设对开门，内容抽屉，可以放置杂物，上方有一平台，镶一个可折叠的镜架，古人用于放置铜镜和化妆用具，今人喜欢其玲珑剔透、光彩熠熠的装饰效果，作为收藏品，放置案头，添了几分富贵。小多宝格是一种可搁置案头或者挂在墙壁上的架子，造型优美，倘若在上面放上主人心爱的紫砂壶、青花瓷等藏品，既悦己，又悦人。

The following chapter includes the clothes rack, the lantern rack, the brush rack, the mirror rack (stand) and small trestle antique display; the clothes rack is used to hang up the clothes in ancient times, commonly with beautiful structures, therefore also functions as a decorative item in the interior; the lantern rack is structured in the single post-style, with slender and elegant appearance; as a daily necessities in ancient times, the delicacy and decorative detail of lantern rack varied due to different economic abilities; the brush rack is more like a miniature of clothes rack, usually being placed on the table as the requisite for calligraphy practicers; the mirror rack is widely used in ancient times, constructed with a cabinet-styled body with two door panels and drawers insides to contain small things and items, and a platform attached above the cabinet embedded with a folding mirror rack on which the ancients could put copper mirror and toiletries; today, the mirror rack is pursued for its delicacy and decorative quality, and is placed on the table as private collection to add a sense of elegance and archaic beauty in the room; the trestle antique display is a rack mostly placed on the table or hung on the wall, with beautiful structure and a practical value to display the owner's beloved dark-red enameled potteries and blue-and-white porcelains.

123

如意云纹衣架

Clothes Rack with *Ruyi*-shaped Cloud Design

133.5×38.5×166.5cm

衣架由白酸枝制成，色泽淡雅温润，用料纤细。

衣架由两块横木做墩子，墩子下挖出两组落地。墩子上立柱，柱前后以站牙抵夹。站牙做如意云纹。两墩之间安由横竖材组成的棂格，使两墩子加固稳定，并可以承托鞋子等物品。

两立柱偏下位置架横枨，横枨之下安如意云头角牙，既起到加固两立柱的结构作用，又可搭放短的衣服等。两立柱最上安搭脑，搭脑两端出头，向上翻转出圆雕灵芝云头纹饰。搭脑与立柱相交处内外两侧，皆安如意云头角牙，内外侧角牙对称排列。搭脑之上安钩，可挂衣物。

此衣架比例修长高挑，异于传统的衣架比例，在使用方式上也有所创新。传统衣架是把衣服搭在衣架的横枨之上。而此件衣架则是安钩，像现代生活方式一样，可以挂衣服。此衣架既是造型的创新，也引入了现代的使用方式。

The clothes rack is made of slim and slender white dalbergia cochinchinensis material with smooth texture and elegant colour.

The rack is constructed with two pedestals installed with posts braced by vertical plates processed in openwork of *Ruyi*-shaped cloud pattern; between two pedestals connects a horizontal lattice work assembled by small straight components to help to fix the base and also to hold shoes or other things.

Between the lower parts of two posts installs a stretcher with corner plates processed in openwork of *Ruyi*-shaped cloud pattern to better fix the structure and also to hang up short clothes; the top parts of posts are installed with a horizontal rail with two ends curved up and carved with ganoderma-shaped cloud patterns; below the corners of the intersection of the rail and two posts installs symmetrical corner plates processed in openwork of *Ruyi*-shaped cloud pattern; along the top horizontal rail installs a row of hooks for hanging clothes.

This clothes rack is built in slender appearance, varied from the structural scale of traditional clothes racks, with some original adaptations. Traditional clothes racks are only built with stretchers for hanging clothes; however, this one comes with hooks as modern clothes racks, introducing an up-to-date design.

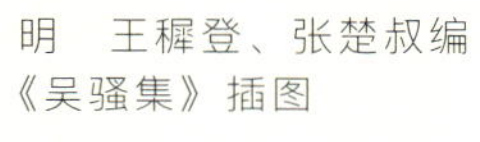

明　王穉登、张楚叔编
《吴骚集》插图

Ming Dynasty (1368-1644)
Illustration of *Collection of Verses Written by Litterateurs of Ming Dynasty*
Edited by Wang Zhideng, Zhang Chushu

124

三兽吞足独梃式灯架

Single Post-styled Lantern Holder with a Design of Swallowing Beast Head

35×35×154cm

灯架以黑料红酸枝制成。

其造型为独梃式，即中间以圆杆状构件为支柱。上方有圆托盘，周沿做成素混面，上下有压边线。承接灯罩，下方装三个挂牙，透雕卷草纹，草叶舒卷自然。独梃下方亦穿托盘，以便与三个腿足合理衔接。三弯腿足，肩部雕为兽首吞足状，腿足下方外翻为卷叶纹，勾卷一个圆珠。下方踩在莲苞上，寓意清雅干净。腿足间有三角形构件，托住自腿足上方穿下的独梃。

灯架是明式家具中少见的品种，多成对陈设，是营造中式室内环境上好的点缀，其造型秀丽而丰富，宛如雕塑，既可装灯泡作灯实用，又可作为摆件放置，客厅、书房、卧室无一不可。

The lantern holder (rack) is made of red dalbergia cochinchinensis with black colour.

It is built in single post-style, with a round stick installed in the middle as the major support structure, and attached with a round disc on the top with cambered edges and two rabbets along top and bottom rims to hold the lantern-shade; below the disc installs three hanging plates carved with openwork of floral patterns; the lower part of the post comes through another disc so as to better connect with three legs; the three-curved legs (cabriole legs) are carved with a design of swallowing beast head on shoulders; three feet are curved up into design of a pearl wrapped by floral leaves; three lotus-shaped foot pads are attached below to imply the meaning of cleanness and peacefulness; a triangular component is installed between legs to fix the post end coming through the disc. Lantern holders are rarely seen among Ming-styled furniture, usually coming in pairs as an ideal decoration in the room; with elegant appearance and various structures, the lantern holder is more like a sculpture, with both practical function as a lighting equipment and also decorative function as an ornament in the living room, the study and bedroom.

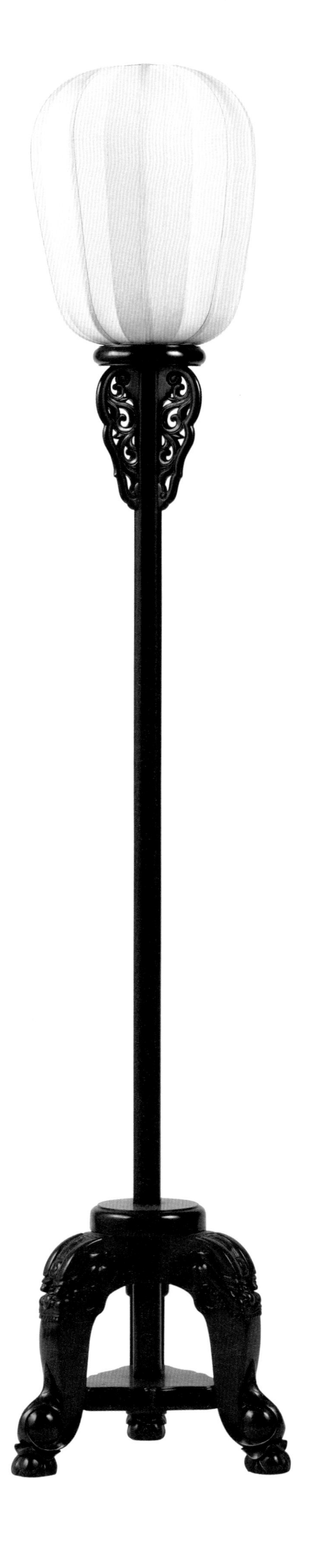

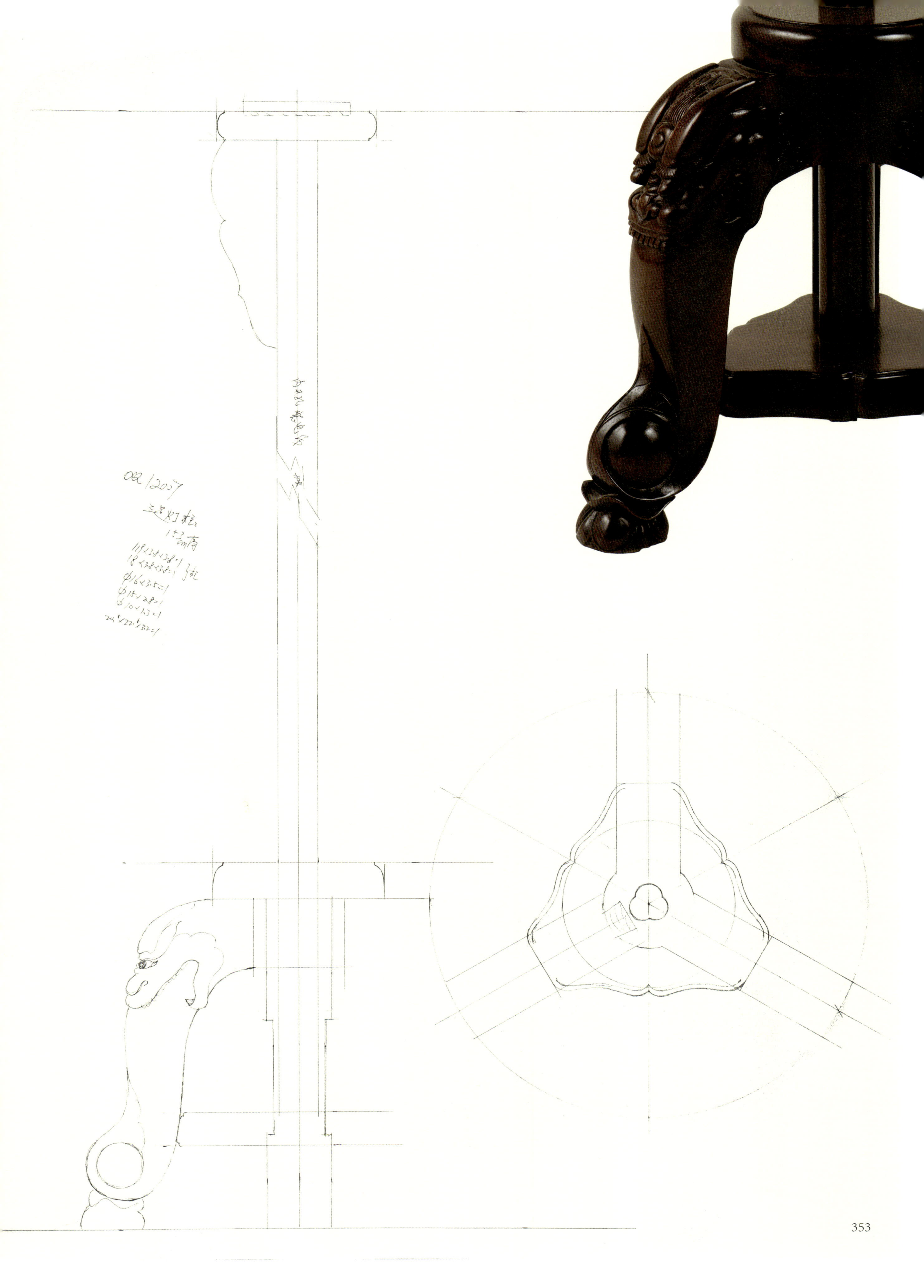

125

凤穿牡丹纹七屏式镜台

Seven-Panel Mirror Stand with Openwork of Phoenixes Flying through Peony Flowers

56×31×86cm

此镜台由黄花梨制成，黄花梨色泽淡雅温润，纹理细腻，麦穗纹清晰。

镜台采用宝座式，即台座之上做七屏风宝座式后背和扶手。后背三扇，中间高，两侧低，搭脑皆中间拱起，两端下垂，又翻转，圆雕龙头。正中一扇中间拱起搭脑正中上安火珠，有双龙戏珠寓意。搭脑之下心板上正中透雕凤穿牡丹纹饰，两侧透雕花鸟纹饰。心板之下又用横枨界出两空间，中起壶门开光，下起壶门亮脚。扶手两扇，依次变矮，亦透雕花鸟纹饰。

台座上前面对开门，门上安圆形合叶和圆形面叶。底座为三弯腿和卷草牙子。

此镜台屏风式，为镜台中等级较高者，在素雅的台座之上做各式雕琢的七屏风式，华丽秀美。

This mirror stand is made of yellow *Huali (Huanghuali)* wood with elegant colour and refined natural veinings.

The stand is built in throne-style, with a thron-styled backrest constructed with seven-panel screen and two armrests; the back panel consists of three leaves arranged in " 山 " style, with hunched top rails decorated with carvings of dragon heads and a firing pearl in the middle, implying the meaning of "double dragons frolicking with a pearl"; below the top rails is installed with three panels carved with openwork of phoenixes flying through peonies in the middle and birds and floral patterns on two sides; below the panels installs a stretcher to divide the below *Kun*-gate apron plates and corner plates; armrests are built with two panels of different height and carved with openwork of birds and floral pattern.

The cabinet-styled pedestal is hinged with two doors inlaid with round latch base and escutcheons, installed with four three-curved legs (cabriole legs) and bottom apron plates decorated with floral bas-relief.

This mirror stand is built in the seven-panel-screen style, absolutely a high-leveled piece among mirror stands, and decorated with various carvings to reveal an elegant and luxurious beauty.

126

折枝花卉纹镜台

Mirror Stand with Floral Bas-relief

42×42×29cm

镜台由紫檀制成，紫檀色泽沉穆，纹理细腻，其上镶嵌的铜饰件色泽浅淡，形成鲜明对比。紫檀适合雕刻，此镜台上满雕各种纹饰，为典型的紫檀工。

镜台上层边框内围支架铜镜的背板，通过后面支架，可以支成斜面，也可以放平，存放在层边框内。背板攒框而成，以长短不同的横竖枨界出三层八格。下层正中一格安荷叶式托，可在格内上下移动，以支直径不同的铜镜。中层方格正中安四角牙，成四簇云头。其余六格接装绦环板，板上浅浮雕折枝梅纹。

镜台上层之下做框架结构，前面对开门，门四面攒边，中装边，上浅浮雕梅枝纹。两门中间安条形面叶，上安吊牌和钮头。门两侧竖框伸出，做木门轴，安在门下横枨挖出的臼窝里。侧面立墙之上亦浅浮雕折枝梅纹。

最下为底座，底座四面攒边，下安壶门牙子和四腿。牙子之上浅浮雕卷草纹饰，腿三弯，于末端外翻。底座四角镶铜包角。

此镜台为标准制式，镜台背板、两门、侧墙和底座牙子之上浅浮雕的梅枝纹饰、卷草纹饰，使得整件镜台精致秀雅，气质不凡。

This mirror stand is made of red sandalwood with dark and solemn colour and refined veinings, inlaid with pale white copper fittings to form a striking contrast in colour and texture. The red sandalwood is an ideal material for carvings and sculpting. This mirror stand is densely decorated with various bas-relief, absolutely a typical craftsmanship applied to such wood.

The top lid is constructed with assembled frames and panels to support the copper mirror; a holder is installed at the back of the lid to form a slope or a flat; the top lid is divided by several rails and struts into eight rectangular grids; the middle grid at the bottom is attached with a movable lotus-shaped supporter to hold copper mirrors of different sizes; the middle grid in the centre is installed with four corner plates carved into cloud-shaped openwork; the other six grids are embedded with decorative panels carved with bas-relief of plum blossoms.

Below the top lid, the stand is built in cabinet style installed with two doors constructed with assembled frames and central panels carved with bas-relief of plum blossoms; rectangular latch bases are inlaid on the front doors with handles and latch pin; the pivots of door panels are fitted in the sockets gouged in the lower stretcher; the side panels are also decorated with bas-relief of plum blossoms.

This base is constructed with assembled frames, *Kun*-gate apron plates carvd with floral bas-relief and four three-curved legs (cabriole legs); four corners of the base are wrapped with copper leaves.

This mirror stand is built in typical structure, with top lid, two doors, side panels and bottom apron plates all carved with bas-relief of plum blossoms and floral patterns to increase its delicacy and elegance.

127

灵芝纹笔架

Brush Rack with Ganoderma Carvings

31×13.5×42cm

笔架黄花梨制成。

罗锅式搭脑，曲线柔和，两端上翘为灵芝头，雕琢生动，打磨仔细，栩栩如生。搭脑上方镂雕双螭捧寿纹，除了吉祥寓意外，造型上又总领整体，起到了画龙点睛的效果。搭脑上纵向植短枨，末端弯曲，可以勾挂毛笔。立柱为委角方形，外侧设简化螭纹挂牙，与搭脑处图案呼应。立柱下方连以双枨，枨间设绦环板，透雕海棠形开光，中段袅起如意头。下枨下方设披水牙板，壶门式，沿边起线，至牙头处勾卷，有草芽努出。立柱两侧设镂雕螭龙纹变体站牙，下承坐墩，几座式，腿足外撇，下方挖壶门亮脚。

此笔架简繁得当，本身就是一件优美的艺术品，且不耽实用。

This brush rack is made of yellow *Huali (Huanghuali)* wood.

It has a hunched top rail with smooth curves and two ends everted and carved into ganodermas which are carefully polished and true to life; above the top rail is carved with image of two dragons holding up the Chinese character of " 寿 " (longevity), implying auspicious meanings and also directing people's focus to the top with its delicate finishing touch; along the top rail installs a row of hooks for hanging writing brushes; the two posts are processed into rectangular shape with dented corners and attached with two hanging plates on two sides to echo the carving above the top rail; between two posts installs double stretchers and embedded with frieze plaque in between carved with openwork of begonia-shaped tracery and a *Ruyi*-shaped cloud in the middle; below the bottom stretcher installs *Kun*-gate apron plates outlined with embossed borderlines and carved with floral design at corners; two posts are braced by vertical plates carved with openwork of dragon patterns and attached with trestle-styled pedestals below; four feet are curved up and installed with apron plates.

This brush rack is built in a well-balanced structure, with both practical and aesthetic values.

128

蝠磬纹瓶形小多宝格

Small Vase-shaped Table Antique Shelf Carved with Design of Bat-shaped Chime Stone

33×13×59cm

小多宝格以紫檀制成，造型别致而优美。

多宝格主体为瓶形，上方设蝠磬纹装饰，为磬形，又似一变体蝠纹，寓意“福庆有余”，表面又有勾云纹、卷云纹、兽面纹和回纹装饰，采用圆雕加浮雕、透雕工艺，雕刻工整而造型优美，宛若为瓶子加设了盖子。

瓶形主体造型矫健，如同将一个圈足屏拉伸，其边缘雕刻有繁密的回纹，又名扯不断纹。中间以横竖板隔成数个大小不一的空间，可以放置大小高低不同的物品，隔板的侧面皆镂空如意云头式开光，如同园林中的花窗，隔而不断，意趣丰富。下部收进为须弥座式足，束腰处浮雕连绵的如意云纹和蕉叶纹，为仿自青铜器的纹饰。其下承以底座，雕刻为繁密的海水纹，如此，宝瓶自海水中涌出，有福海献宝之意。

This small table antique shelf is made of red sandalwood, with beautiful structure.

The shelf is built in vase-shaped appearance, decorated with a delicate carving of bat-shaped chime stone, as a variation of bat pattern, implying the blessing of "having superfluous happieness", with its surface carved with cloud-shaped patterns, beast face motif and fretwork by various craftsmanships including circular engravure, openwork and bas-relief, as if the vase-shaped body is covered with a lid; the body of the shelf is built in vase-shape with its outer frames densely decorated with bas-relief of fretwork; the middle area is divided by compartment panels into several spaces of different sizes for diplaying antique objects of various sizes; the compartment panels are carved with openwork of *Ruyi*-shaped cloud pattern like the traceries in the Chinese gardens; the bottom is attached with a sumeru-seat-styled pedestal with girdled waist carved with continuous bas-relief of *Ruyi*-shaped cloud and banana-leaf motif as an imitation of the pattern on bronze ware; the bottom part is densely carved with raging waves to imply the image of "treasure vase emerging from the ocean".

129

切角长方形月窗式可悬挂小多宝格

Small Rectangular Hanging Antique Shelf with Design of Window Tracery and Dented Corners

40×5×50cm

小多宝格紫檀制成，切角长方形，框为八面攒边而成。攒框之内，向内攒花板，花板略高起。花板再向内攒花板，花板平齐。如此形成一个略高起的花框。花板上皆透雕折枝花纹饰。上部再安牙子，牙子上透雕拐子蝙蝠纹饰。

画框之内，拉伸空间，为可搁置物品的小多宝格。画框设计成室内一角，月窗露出一角，透雕卷草纹饰和拐子纹饰的花板清晰可辨。圆雕床帐，帐幔的翻折动感，和卷帐的绳索亦隐约可见。月窗之下，安小格如同桌案。

此挂屏设计了方形折角花窗之内露出的月窗一角，形象生动，喜剧情节隐含其中。

此小多宝格为创新设计，非传统制式，独出新意。

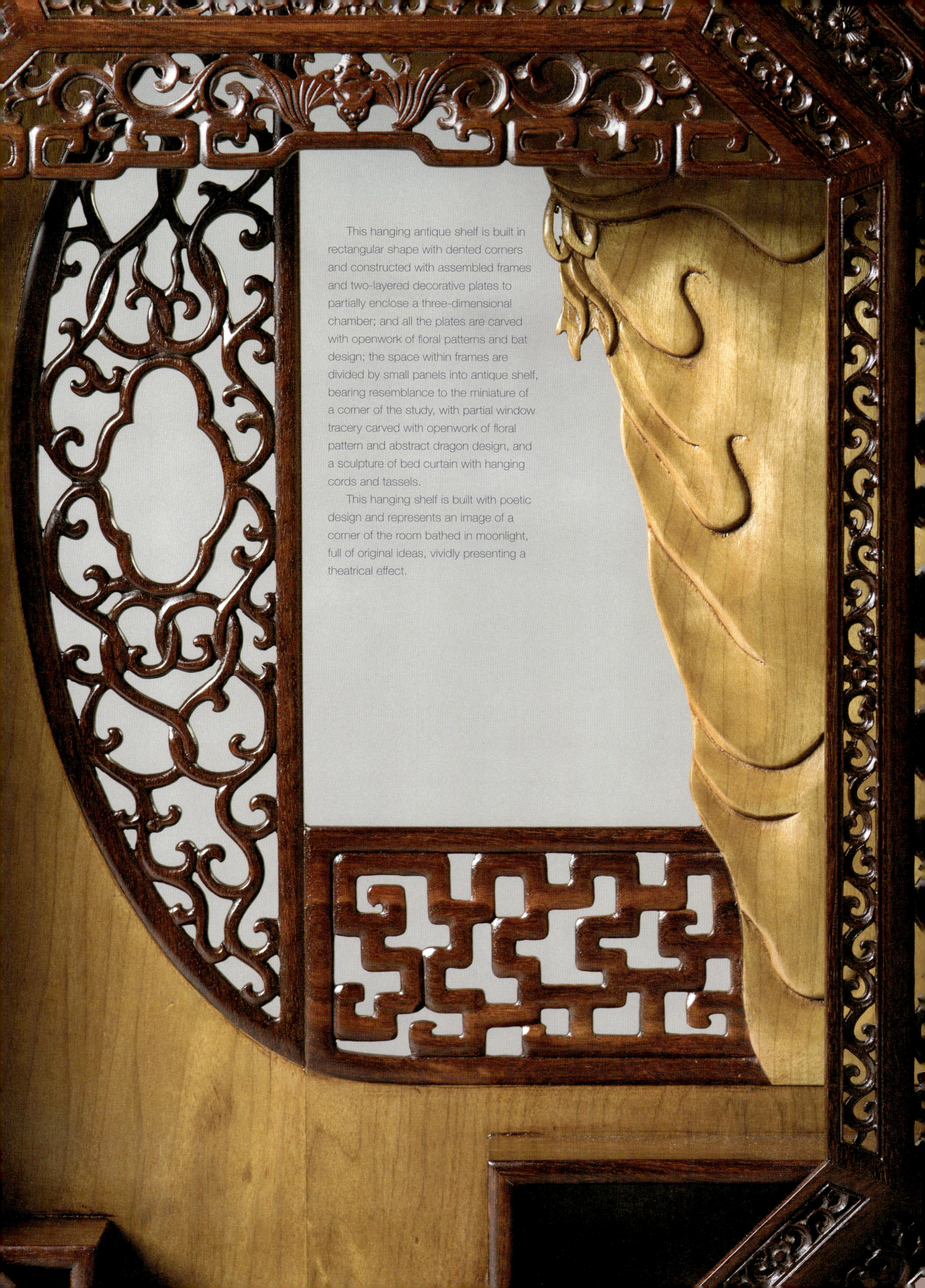

This hanging antique shelf is built in rectangular shape with dented corners and constructed with assembled frames and two-layered decorative plates to partially enclose a three-dimensional chamber; and all the plates are carved with openwork of floral patterns and bat design; the space within frames are divided by small panels into antique shelf, bearing resemblance to the miniature of a corner of the study, with partial window tracery carved with openwork of floral pattern and abstract dragon design, and a sculpture of bed curtain with hanging cords and tassels.

This hanging shelf is built with poetic design and represents an image of a corner of the room bathed in moonlight, full of original ideas, vividly presenting a theatrical effect.

文玩

Antique Objects

晚明文人文震亨《长物志·器具》记："古人制器尚用，不惜所费，故制作极备。"文玩之属，是最不必要的长物，然人之雅致、闲情，却能在这些长物中得以寄托，它们或精致，或典雅，或气宇轩昂，或小巧玲珑，搁置案头，得时抚摸，一切烦恼顿消；得时瞩目，神游物外，精神世界又得大的丰富。文玩者，一在"文"，言其为读书人的用具，是文人审美体系里的东西，和书卷、笔墨纸砚一样，都是文人生活中必不可少的一部分，不必多精贵，但一定是雅致的，充满了人文关怀的；二在"玩"，玩具也，不必太认真，放松品鉴，逸兴四飞，就会发现这其中的真韵、真情、真意。

According to *Superfluous Things* written by Wen Zhenheng in the late Ming Dynasty, it says, "the ancient people always aimed to make durable utensils and wares, therefore, all the best resources and labours were devoted into the production."As for the antique objects, they seemed to be the least needed things during daily life, however, they could reflect the owner's aesthetic taste and cultivated manners. These objects are of delicacy or elegance; with grand solemnity or exquisite beauty; they can be placed on the table or caressed in palms; whenever you see them, it expels all the anxieties and worries, and takes you away on a spiritual trip. Its name, literati antiques, implies its true essences: literati, indicates they are mostly collected and appreciated by the educated people, and belong to their aesthetic system, like books, writing brushes, ink sticks, papers, and inkstones, as an indispensable part in literati's daily lives, therefore, they must be elegant and full of humanistic spirit; antiques, indicates that they are merely objects or toys for adults; people needn't take them too seriously; with relaxed feelings and unrestrained thoughts, people can finally see the true meaning in it.

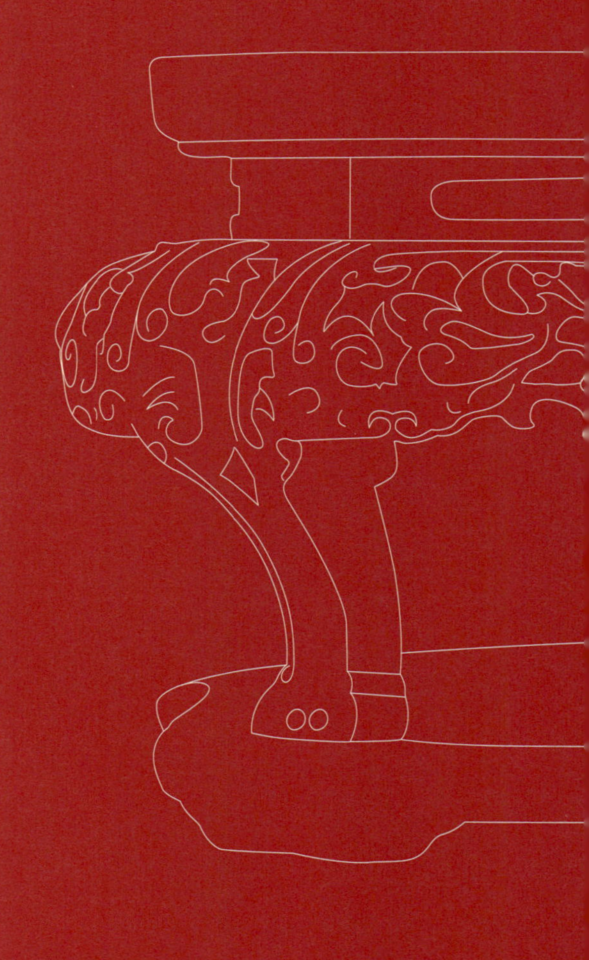

130

如意云纹板足案上几

Small Trestle Table with Panel Feet Carved with Openwork of *Ruyi*-shaped Cloud

48×20×15cm

案上几由白酸枝制成。造型颇简，主要由三块板做成，配合牙板、翘头，即成一器，但各处细节变化多端，非常富有装饰性。

其翘头不做尖角，圆润可爱，素冰盘沿，下端隐有压边线。两板为足，用较厚的材料挖成往外弧突状，下端又往外撇出，变化微妙，侧面看为三弯腿外翻马蹄式样。板足侧面又锼雕如意云纹开光，大小适度，简练可爱，打破独板带来的沉闷感。细微之处在于板足从侧面看，自上往下渐渐挓出，至下端又自然收敛，变化自然，视觉上更觉敦厚可爱。牙板挖去甚多，以便与板足曲线柔婉相接。

此类造型的翘头小几多出在苏北地区，作放置古玩或者小型盆景之用。安思远旧藏有相同制式者，《维扬明式家具》亦收录有与此件尺寸造型几乎相同的实例。

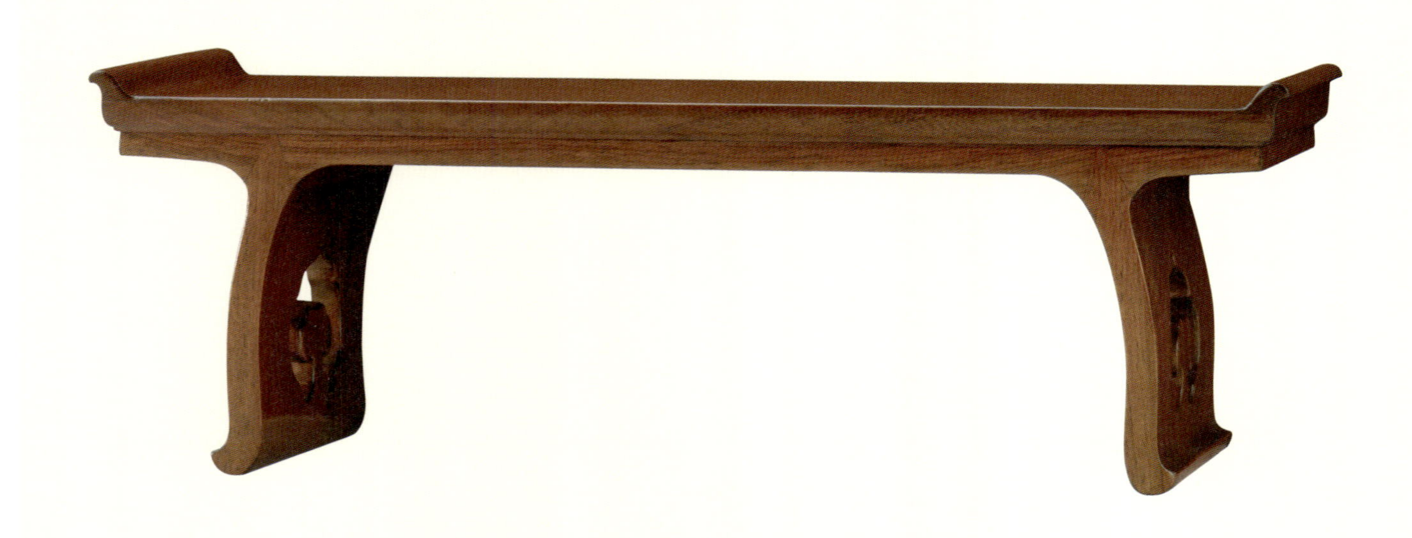

The small trestle table is made of white dalbergia cochinchinensis, with simple structure mainly constructed with three panels, along with decorative apron plates and everted flanges, with many variations in details.

The two everted flanges are polished in rounded silhouette; and the top board is with edges narrowed down widthways (also called ice-plate edge) and a rabbet along the bottom rim; below the top board installs two panels as legs with slight curves resembling to the three-curved legs with hoof feet; the panel legs are carved with openwork of *Ruyi*-shaped cloud to balance the simplicity and dullness; from the side view, two splayed legs are processed into subtle and beautiful curves; the unadorned apron plate below the top board is planed thinner to better connect with the legs.

The trestle table with such structure mostly appears in the northern areas of Jiangsu Province, and is used to display literati antiques or small potted landscape; Robert Hatfield Ellsworth once had a similar one; the book *Ming-style Furniture in Weiyang Area* also includes an example with the same size and structure.

131

缠枝莲纹菱花形带托泥小几

Lotus-shaped Small Stand Carved with Openwork of Intertwined Lotus Motif (with Foot Rails)

30×19×19cm

此小几由紫檀制成，紫檀色泽沉穆，纹理细腻。

几面为菱花式，面心攒边，面心平装薄板，边抹做混面压边线线脚处理。

几面下安束腰，束腰做长鱼门洞式开光。束腰之下安三弯腿和牙子。三弯腿按菱花内收委角的曲线翻转，上透雕卷草纹饰，下收云头足。牙子上透雕缠枝莲纹。牙子和腿子以雕琢连接。

腿足之下安托泥，托泥亦做菱花式，小收小足。

此几造型别致，菱花式贯穿几面、束腰、腿足和托泥，形成一致的造型语言，使得整体造型独特雅致，气质秀雅。

This small stand is made of red sandalwood, with dark and solemn colour and refined veinings.

It has a lotus-shaped top board with assembled frames, a flat central panel, and cambered edges with a rabbet along the bottom rim.

Below the top board installs a girdled waist carved with a pair of hollowed tiny slim long-fish-doorway windows on each side; below the waist connects three-curved legs (cabriole legs) and apron plates; legs are carved with floral openwork at shoulders and cloud-shaped design at four feet; apron plates are carved with openwork of interlocking lotus flowers, smoothly connecting with four legs' shoulders.

A lotus-shaped foot tray is attached below four feet, with small foot pads at the bottom.

This small stand is with delicate structure and decorated with unified lotus flower theme through top board, girdled waist, legs, feet and foot tray, giving this small piece of furniture a unique delicacy and elegance.

132

高束腰菱花形三连式小几

Joint Triplet Lotus-shaped Small Stand with High Girdled Waist

39.5×14.5×12cm

此三连几由紫檀制成，紫檀色泽沉穆，纹理细腻，适合雕刻。

此三连几造型独特，由一大两小三个香几连接牙子而成一整体。三个香几造型基本一致，只在体量、细节上略有不同。

几面为菱花式，下承束腰。束腰较高，上开海棠式开光，开光之内透雕各种不同纹饰。束腰之下安三弯腿和牙子，三弯腿上部膨出，然后向内收，中间翻出卷草纹，于末端外翻如意云头收尾。束腰S形曲线夸张，为整料挖出，视觉效果突出。牙子做云头曲线，边缘起阳线勾勒。S形腿收敛最窄处安一圈托泥，托泥纤细。

三单体香几之下各承一底座，三底座造型基本一致，在相邻的牙子腿足之处连接一体。底座为束腰式，下安拐子形腿足和牙子。

此三连几造型独特，非常规制式，雕琢精细，工艺精湛。

This small stand is made of red sandalwood, with dark and solemn colour and refined veinings, an ideal material for carving and sculpting.

It is built in quite a distinct structure, with one big and two small stands jointed by decorative plates in between; three stands are of the same appearance, only varied in size and small details.

They have three lotus-shaped top board, with a high girdled waist carved with begonia-shaped windows and different patterns within; below the waists connects three-curved legs (cabriole legs) with dramatically bulged shoulders, floral carvings in the middle and cloud-shaped design at curved feet; the beautifully curved s-shaped girdled waists are processed out of an one-piece material; apron plates are carved with cloud-shaped openwork and outlined with embossed borderlines; slim foot rails are installed at the curved parts of legs.

Each stand is attached with a pedestal of similar structure which are connected by apron plates; the pedestals are with girdled waists, geometric feet and apron plates carved with openworks.

This joint triplet stand is built in quite a unique structure, with original design and delicate carvings and decorations, as a demonstration of highly skilled craftsmanship.

133

弦纹箸瓶

Chopstick Vase with An Embossed Chord Motif at Rim

4.3×11.2cm

箸瓶以红酸枝旋制而成，纹路明显，黄褐色，沉稳内敛，质感甚佳。

形若双陆棋子，口沿有凸弦纹一道，内侧为柔和的唇口，长颈，上腹部渐凸，越往下凸出越多，形成饱满的圆弧腹，下端内收，并翻出阳线一周为弦纹。

箸瓶本是古人炉、瓶、盒三事中的瓶，用来插焚香时用的筷子，故名。传世箸瓶造型多别发机杼，加工细致，选材上佳，是独特的一种文玩，今已经成为案头陈设或者把玩的小器具，是否插箸已不重要，心悦雅玩而已。

This chopstick vase is made of red dalbergia cochinchinensis, with clear veinings, light brown colour and refined texture.

This vase looks like a chess piece, with an embossed chord motif at rim; the vase is processed to have a smooth silhouette, from the well-polished rim, slender neck, to bulged belly, forming a beautiful curve; the bottom is also decorated with an embossed chord motif.

The chopstick vase is originally used in ancient incense ritual for containing incense chopsticks, hence the name, along with the incense burner and incense case; the existing ancient chopstick vases, as unique literati antiques, are mostly built in original structures, processed with intensive craftsmanship, and made of highly selected material. At present, it has become a small ornament placed on the table (without chopsticks).

134

瓜棱式箸瓶

Chopstick Vase with Melon-shaped Section

4×11.8cm

箸瓶以红酸枝制成，木质细腻，色调沉稳，木纹优美。

菱花形口，长颈，口部微侈后渐细，宛若鹅颈，腹部自然顺畅地突出又回收落地，形如蒜头，菱花形自口部一顺而下，随形体而起伏，形成瓜棱效果。

此瓶虽未刻意雕琢纹饰，但形体优美，菱花形截面附有自然有韵律的装饰，亦是一件上佳的文玩器具。

This chopstick vase is made of red dalbergia cochinchinensis, with refined texture, solemn colour and beautiful natural veinings.

It has a lotus-shaped mouth and slender neck which is slightly narrowed down as a swan neck; the belly is smoothly bulged in the middle as a garlic bulb; the whole body is carved with ridges from the lotus-shaped mouth to the bottom, giving the vase a melon-shaped section.

Although the vase is not heavily carved or decorated, its curved silhouette and ridges are of great elegance and natural rhythm.

135

云龙纹带底座香筒

Pedestal Incense Canister Carved with Openwork of Cloud and Dragons

12×12×42cm

香筒由紫檀制成，紫檀色泽沉穆，纹理细腻，适合雕刻。

香筒体形修长，上盖顶盖，下收底座，中为圆柱形筒身。顶盖圆口穹顶，上安莲蓬状把手，下为覆莲造型，盖身透雕云龙纹，与筒身以子母口扣合。

筒身口沿外浅浮雕莲瓣纹，下承束腰，束腰上下做仰覆莲，中为圆柱形，上透雕云龙纹，龙身翻腾于祥云之中，云气氤氲，与香炉焚香功用相合。柱体下再收束腰，束腰上下做仰覆莲。下接三弯腿和牙子，三弯腿和牙子一木连做，上浅浮雕纹饰。

此香筒造型特别，雕刻精美，工艺精湛，为精工之作。

The incense canister is made of red sandalwood, with dark and solemn colour and refined veinings, an ideal material for carving and sculpting.

The incense canister is built in a slender cylindrical structure, with a lid above and a pedestal below. The lid is with a circular rim and a handle on the top carved into lotus seedpod; the lid is decorated with openwork of cloud and dragons, and can be fastened on the canister body.

The body is carved with bas-relief of upturned and downturned lotus petals next to the rim with a girdled waist in the middle; the major part is built in cylindrical structure, carved with openwork of cloud and dragons; the image of dragons flying through the cloud echoes its function of burning incense; below the canister body installs the other girdled waist between the same upturned and downturned lotus petals; the pedestal below is attached with three-curved legs (cabriole legs) and apron plates decorated with bas-relief of cloud pattern.

The incense canister is built in a quite distinct structure, with exquisite carvings and highly skilled craftsmanship.

136

黄杨木瘦骨罗汉摆件

Boxwood Sculpture: Skinny Arhat

15×12×20cm

此圆雕作品由黄杨木制成，黄杨颜色淡黄，因陈年氧化，略有包浆，而成褐黄色，包浆莹润，纹理细腻。

此雕刻为瘦骨罗汉，骨瘦如柴，筋骨暴露，精神矍铄。罗汉坐于树叶做成的蒲团之上，一腿盘腿，一腿屈立，左手置于左膝盖之上，右胳膊搭于左胳膊之上，左手持念珠一串。罗汉身形瘦削，脊肋历历可数，罗汉面部慈善，宽鼻阔耳，双目半阖，唇角上扬，作静思状。

瘦骨罗汉又称为“雪山大士”，是释迦牟尼的别称，佛祖释迦牟尼，因见众生疾苦，随入雪山苦行，终在痛苦中得道，参透成佛。

This circular engravure is made of boxwood, with refined veinings and a thin layer of oxidation film due to its aged history, turning the original light yellow colour of the boxwood into light brown.

This small sculpture is modeled on the Skinny Arhat, with contours of bones clearly looming behind the skin, and with a vigorous expression depicted on his face; The arhat sits on a cushion made of leaves, with one leg bended inward, the other crooked up, and his left hand holding a string of beads; the arhat is extremely thin and scrawny, with each ribs clearly carved out along his belly; he has a merciful expression, with wide nose and long earlobes, half closed eyes and a big smile, posing for meditation.

The Skinny Arhat, also called Master of Snow Mountain, is an alternative name for Sakyamuni who, after witnessed the pains and sorrows suffered by all the beings, decided to withdraw into the snow mountain and to live a ascetic life, then finally awakened to the enlightenment and become the Buddha after he went through all the sufferings and trials.

137

鹿负灵芝摆件

Sculpture: Deer Carrying A Bunch of Ganodermas

20×10×20cm

紫檀圆雕一鹿，体形壮硕，饱满圆润，四肢轻巧，似在行进。

鹿高颈细脖，鹿嘴上翘，鹿眼清纯，眉目含情，作回首状。鹿角粗硕有力，分叉匀称，可见鹿龄不小，有能力承担重任。鹿背之上背负书卷和灵芝，书卷以绳缠起，灵芝似生发之状，挺拔向上。鹿回首似在查看背负的宝物，以做保护之意。

圆雕鹿行进于云气之上，底座雕刻为氤氲云气，挖出承鹿四足之处，巧妙稳定。鹿与底座同时设计搭配，上下协调统一。

鹿与禄字谐音，象征吉祥长寿和升官之意，灵芝又是仙草，多为仙鹤和鹿来保护，有灵鹿衔芝的纹饰。此圆雕作品为灵鹿回首保护背上的灵芝，寓意颇为相似。

This deer sculpture is carved with a sturdy body, slim and slender legs, as if it was walking in the mountain.

The deer carving is posed to look back, with a long and lithe neck, upturned lips, and adorable eyes. Its antlers are depicted hard and stout, evenly splayed, indicating its maturity as well as the capability to carry the weight; it is carrying a bundle of books and a branch of ganodermas on the back; the deer seems to turn its head back to look at the treasure it is carrying, showing a sense of protectiveness.

The deer seems to walk above the cloud as the pedestal is carved with abstract cloud design; four hooves touch on the cloud-patterned ground to maintain the steadieness of the whole sculpture, showcasing a unified style and context.

The character "鹿" (deer), pronounced as *Lu* in Chinese, is an homophone for "禄 (*Lu*)", symbolizing having promotions and a great fortune; and ganoderma is a rare herb for prolonging people's lifespan, usually protected by cranes or deers, hence the image of fairy deer holding a branch of ganoderma in its mouth; this sculpture depicts a deer looking back at the ganodermas it carries on its back, implying the similar auspicious meanings.

138

童子牧牛摆件

Sculpture: Shepherd Boy with a Buffalo

45×25×24cm

此圆雕作品紫檀制成，雕一南方水牛卧于枝草之上，作休憩状。水牛前肢伏地，后腿顺于身前，露出宽阔的牛蹄。水牛头上一对牛角粗硕有力，向头后弯曲。牛角前面两眼鼓凸明显，方口阔鼻，鼻上穿环，系绳，绳牵于牛背的牧童手上。水牛身上的牛毛丝丝清晰可见。水牛回首，与牧童呼应。

牧童身着宽服，背斗笠，头梳两髻，面带笑容，回首远看。牧童左手持书卷，右手牵着绳。似是身后有事，惹得童子和水牛同时回首观看，圆雕捕捉到这一传神瞬间，记录下来，生动形象。

水牛下承底座，底座为虬枝搭成，正合水牛身量，量体裁衣，正成一套。

此件圆雕作品雕琢细腻生动，具有浓郁的乡村野趣。

The sculpture depicts a southern buffalo lying on the grassland to have a rest.

The buffalo, with its forelegs crouched and hind legs bended forward, showing its wide hooves; a pair of sturdy and strong horns are curved backward on its head; it has big bulging eyes, a square mouth and a wide nose strung with rings tied up to the rein held by the shepherd boy sitting on its back; its hair is clearly depicted; the buffalo is looking back to respond to the shepherd boy.

The shepherd boy is wearing loose clothes and a bamboo hat on his back, with his hair worn in two buns; he is smiling while looking for into the distance; he holds a book in the left hand, seizes the rein in right hand; it seems that something happened at their back which draws their attention, and the sculpture perfectly catches the moment.

The buffalo is attached to a pedestal carved into branches and leaves, with well-balanced proportion.

This sculpture is with meticulous details, represents a lively scene in the countryside.

139

衔灵芝螭虎摆件

Sculpture: A Mythical Creature
(with Features of Tiger and Dragon)
Gripping A Ganoderma

36×13×5cm

此作品紫檀制成，圆雕一卧形螭虎，身体匍匐于地，虎头侧倾，两眼鼓凸，阔鼻长嘴，警觉竖耳。嘴紧闭，口衔一大朵灵芝，螭虎一前爪按住灵芝头部，一前爪抓住灵芝根部，似捕获猎物状，后爪则稳稳伏地，做蹬地状，虎尾粗硕，缠绕于身后，做如意状。螭虎动作形象生动，憨态可掬。

此圆雕整体比例修长矮扁，应是置于书房书案之上，可作镇，可置笔，伴读书写字之用。

This sculpture depicts a couching mythical creature (with features of tiger and dragon), with a pair of bulging eyes, wide nose, big mouth and erected ears, crawling on the ground, with its head tilting, mouth closed to grip a branch of ganoderma, its left forehand pinning down the head of the ganoderma and right forehand seizing its root, the hindlegs crouching on the ground, and its strong and thick tail coiled in *Ruyi*-shape. The creature is carved in lively and adorable appearance.

This sculpture is carved in short and slender form, possibly used as paperweight or brush rack.

140

吉祥如意葫芦摆件（一对）

Sculpture: Auspicious Gourd (a Pair)

18×18×44cm

此作品由紫檀制成，紫檀色泽深穆稳重，纹理细腻紧致，特别适合雕刻，此圆雕工艺正是材尽其用。

圆雕一饱满葫芦，葫芦体形修长，颈部高挑，气质不凡。葫芦周身以透雕古币锦地，锦地之上浮雕缠枝葫芦枝蔓，葫芦枝叶繁茂，蜿蜒交错，细长蜷曲的葫芦须缠绕枝叶之上，生动活泼。葫芦枝蔓之间间或几只飞蛾嬉戏其中。葫芦颈上枝干渐粗，穿枝过梗，以最粗枝干收尾。

葫芦上下圆体之上开圆形开光，开光之上镶嵌圆形点。开光之内做六边形锦地，上镶嵌“吉”“祥”“如”“意”字样。

葫芦之下安底座，底座为束腰式，束腰之下安三弯腿和牙子。

此圆雕作品比例修长，造型与雕饰相宜，雕工精湛，精益求精。

The sculpture is made of red sandalwood, with dark and solemn colour and refined veinings, an ideal material for carving and sculpting.

The gourd sculpture is built in elegant and slender form with a long neck; it is covered by openwork of ancient coin design above which is carved with bas-relief of gourds and vines, winding around the gourd's body with curved branches and tendrils, flourishing leaves, and several moths flying among them; the branches grow stronger as they run up to the neck of the gourd and end with large stalks at the top.

Four round plaques are embedded on gourd's body with ivory-inlaid inscriptions of Chinese characters on the ground carved with hexagon openwork, saying "吉"and "祥" ("如""意"on the other side), expressing an auspicious meaning of "everything going well as you wished".

Below the gourd installs a pedestal, with girdled waist, three-curved legs (cabriole legs), and apron plates.

This sculpture is carved in slender form, with well-balanced carvings and structure, showcasing a highly skilled craftsmanship.

141

西番莲纹带毗卢帽方宫灯

Palace Lantern Carved in Monk's Hat-sytle with Openwork of Passiflora Patterns

29×29×64cm

宫灯为红酸枝和黄杨木混作而成，黑黄两色相间，配合清新的折枝花卉图画，颇显艳丽。

宫灯顶为毗卢帽式，透雕西番莲纹，花头丰硕，花叶舒卷繁密，如同盛开的莲花，其下承盘式构件，一周为黄杨木拐子纹围栏。又下若束腰收进，为黄杨木透雕拐子纹绦环板。灯身方形，上有盝顶，四角设悬空的柱状构件，往上出头，如同柱头。周沿装透雕西番莲纹花牙。灯身正中四柱，镶装没骨花鸟画。

灯身下方承以方尊式构件，上下侈口而中腰鼓出，鼓出部分以黄杨木透雕西番莲纹而成。尊式构件下承底座，底座四周有拐子纹围栏，做成冰盘沿，下带束腰，腿足和牙板一木制成，浮雕圆润的卷云纹装饰。

宫灯富丽堂皇，制作精巧，是上好的工艺品。

This palace lantern is made of red dalbergia cochinchinensis and boxwood, with strikingly contrasting colours of black and yellow, and inlaid with elegant literati paintings of flowers and birds, presenting a sense of gorgeous luxury.

The top head is carved in monk's hat-style with openwork of passiflora patterns of fully bloomed flower and curled and interlocked branches and leaves resembling lotus flowers, and attached to a disc-shaped component enclosed by boxwood balustrade fence carved with openwork of abstract dragon design; the lantern body is built in rectangular shape with a truncated pyramid top attached with column-styled hanging posts at four corners (tops appear above the truncated pyramid top); between the posts installs apron plates decorated with openwork of passiflora pattern; the body is inlaid with literati paintings of flowers and birds.

Below the lantern body installs a *Zun*-styled (wine vessel used in ancient times) component with flared top and bottom and a bulging waist made of boxwood and carved with openwork of passiflora pattern; the base is attached below and enclosed by boxwood balustrade fence carved with openwork of abstract dragon design, with edges narrowed down widthways (also called ice-plate edge) and a girdled waist; legs and apron plates are carved out of the same material, with bas-relief of cloud pattern.

This palace lantern is built in opulent luxury, with delicate craftsmanship.

142

人物故事图盝顶六方亭式大宫灯

Hexagoanl Pavilion-stypled Giant Palace Lantern
Carved with Six Scenes of Chinese Drama Plays

133×133×220cm

此宫灯由紫檀制成，紫檀色泽沉穆，稳重大气。

宫灯六方亭式造型，由宫帽、宫身和底座组成。

宫帽较宫身向外延，如垂花门式做法，上圆雕龙纹，动感十足。上接盝顶，顶上安宝珠，上圆雕云龙纹。宫身安六立柱，两柱之间装花板，花板上透雕飞天纹饰，正中做异形开光，开光内嵌黄杨，做戏剧故事纹饰，六面戏剧各不相同。柜身之下安花板一圈，花板上透雕加镶嵌，上圆雕龙纹。此栏杆的处理方法与宫帽相似，上下呼应，成一体。

底座上先围栏杆一圈，栏杆上透雕缠枝花卉纹饰。底座六面攒边，收束腰，束腰上浅浮雕卷草纹饰。束腰之下安壸门牙子和三弯腿，三弯腿上部雕兽面纹，下以兽足收尾。底座之下再踩罗锅式托泥，下再安小足。

此宫灯使用圆雕、透雕、浅浮雕、镶嵌等工艺，整体雕琢满密，繁复精致。

This palace lantern is made of red sandalwood, with dark and solemn colour.

It is built in hexagon-style, with lantern hat, body and pedestal.

The lantern hat stretches out further than the body, and is decorated with hanging pillars sculpted into dragons; the hat is built in tent-styled structure with a treasure pearl on the top carved with cloud and dragon design; the body is constructed with six posts and embedded with decorative panels in between, with openwork of flying Apsaras; each panel is inlaid with boxwood chips to depict different famous scenes from Chinese drama plays; below the body installs a circle of panels carved with openwork patterns and embedded with decorative panels carved with dragon patterns; the balustrade fence is processed in the same style of the lantern hat to maintain a unified theme.

The pedestal is constructed with a circle of balustrade fence carved with openwork of floral patterns; the hexagonal base board is braced by assembled frames and attached to a girdled waist decorated with bas-relief of floral patterns; below the waist installs *Kun*-gate apron plates and three-curved legs (cabriole legs) which are carved with beast-face pattern on the shoulders and beast-foot pattern at foot tips; below the base installs hunched foot rails and small foot pads.

This palace lantern is applied with various carving craftsmanship, including circular engravure, openwork, bas-relief and inlay, to make the whole piece densely covered by luxurious and intricate decorations.

143

回纹小卷几

Fretwork-styled Small Table Shelf

34×20×12.5cm

卷几以紫檀制成。

以薄板一块为面，竖两块为腿足，下方又以板攒成勾卷的回纹，内侧连以板，即成卷几，如同一个倒“凹”字。几面平素，几的前后侧面均打洼线，面下前后方设镂空拐子纹花牙，如同悬挂帐幔，具浓厚的装饰效果。板足外侧设银锭形开光，内透雕双蝠纹、卷叶纹，双蝠为龙首，造型威严，寓意吉祥。下方连接性构件，并非屏板，而是侧面略成罗锅状，为简化的变体方如意头形，面上浮雕西番莲纹装饰。

The table shelf is made of red sandalwood. It is constructed with a horizontal panel as the table top and two vertical panels as the legs and feet, decorated with assembled apron plates carved with abstract dragon pattern; the space below is divided by small panels to form a silhouette of an upturned " 凹 "shape; the table top is unadorned and all the edges on two sides are gouged out a round low-lying groove; apron plates decorated with openwork patterns hang below the top like a curtain, which is full of decorative aesthetics; two side panels are carved with bas-relief tracery and openwork of double-bat and floral pattern within to imply an auspicious blessing; the connecting component below is processed in hunched style, carved with bas-relief of passiflora pattern.

144

镶大理石叠落式小几

Joint Multi-Layered Small Table Shelf
Embedded with Marble Plaques

48×18×21cm

小几以紫檀制成，镶嵌大理石为几面。

叠落式即指两个连为一体的几面上下错落的状态，这种造型使得空间变化更为丰富，其上搁置物品高低错落，颇得意趣。几面均攒框装大理石，冰盘沿甚简，束腰打洼，牙板上下两层，上层光素，下层则为透雕的卷草纹，形成券口牙板，正中间透雕倒垂蝠纹，寓意“福到”。腿足为回纹般勾卷的样式，其中两几靠近一侧的腿均歧出拐子纹，从而连为一体。高几一侧的勾卷腿足末端，雕作一个回首的螭龙。两几侧面透雕蝠磬纹。两几下方又做成一周勾转起伏的拐子纹，如同底托，联为一体。

此几设计经典而制作高超，颇富装饰意趣，放置案头，搁置珍玩正佳。

The table shelf is made of red sandalwood, and embedded with marble slabs as its tops.

The Joint multi-layered style refers to the structure with two stands of varied heights connected on one pedestal, which can increase spatial variation and add more fun and charm to the displayed antiques. The stand tops are constructed with assembled frames and embedded with marble slabs, with edges narrowed down widthways (also called ice-plate edge) and a gouged out girdled waist; below the tops installs double-layered apron plates: the upper layer is unadorned; the lower layer is carved with openwork of floral pattern and an upturned bat in the middle to imply the blessing of "bringing in the happieness"; legs and feet are carved into continuous fretwork and abstract dragon pattern to link the two stands together; the apron plates installed on the high stand are decorated with carvings of dragon heads at the tips of the fretwork; and two sides are carved with openwork of bat patterns.

This small table is built with classic design and highly skilled craftsmanship, full of decorative aesthetics. It can be placed on the table for displaying small literati antiques.

微缩家具

Miniature Funiture

此缩微家具组合，琳琅满目，有六扇折屏一具，圆角柜两对，圈椅一对和单只，马扎两只，南官帽椅三只，北官帽椅一只，杌凳一只，脸盆架一具。造型多为简洁静雅的明式风格，比例合宜，用料讲究，所有器物皆按照原有的榫卯结构原比例缩小而成。

This set of ministure furniture is of various kinds and styles, including a six-panel folding screen, two pairs of round-cornered cabinets, three horseshoe-back armchairs, and two different folding stools, three yoke (official's hat)-back armchairs, one yoke (official's hat)-back armchair with four protruding ends, one stool, and one basin rack (stand), are mostly built in simple and elegant Ming-styled structures, with well-balanced proportion and refined material. All the mortise-and-tenon structures and details are referred from the full-sized actual furniture.

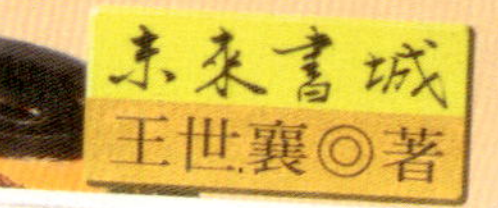

東來書城
王世襄◎著

大象人物聚焦书系
王世襄
找一片自己的天地
李辉 著
大象出版社

北京鴿哨
王世襄 编著
BEIJING GESHAO
BEIJING GESHAO

文化篇

Culture

王世襄最有温度的旧藏

生前自用的四千册图书整体拍卖

程香

朋友圈每天都在被各种捷报和“网红”拍品刷屏，时不时地也会有亿元纪录夺人眼球。在股市和房地产行情持续走低的市场环境下，艺术品成为投资的不二选择，有些热门专场，新老藏家摩拳擦掌争抢涌入，一“座”难求。

而在这些所有的喧嚣之外，今年春拍最打动我、最牵动我心的，是一组成交价仅仅只有 270 万元（含佣金为 310.5 万元）的书籍。

王世襄最有温度的旧藏

这组书籍来自王世襄。

不分新旧，一共 4000 册，以编号 4689 的标的物“王世襄袁荃猷藏书十柜”，一次性出现在中国嘉德“玩物适情——名家收藏集珍”专场。

自 2009 年冬王世襄先生仙逝后，他的一生珍藏，由后人陆续释向市场，从家具到古琴、佛像、香炉、古籍善本、竹刻、鸽哨等，凡“王世襄旧藏”一经出现，必能引起轰动，名家效应可见一斑。

到 2019 年，经过十年的渐渐释出，王世襄生前旧藏已散至天涯各处，这批图书，恐怕是最后一批遗珍。与往年 1.15 亿元的“大圣遗音”伏羲式琴、5520 万元的“明周制鱼龙海兽紫檀笔筒”相比，这十柜旧书显得毫不起眼，被淹没在今年热闹的春拍大潮中，并未引起任何媒体的关注。

王世襄生前自用的 4000 册图书，连书带柜整体上拍

6 月 3 日晚上八点，中国嘉德“玩物适情——名家收藏集珍”拍卖现场人声鼎沸，座无虚席，后半场大多是王世襄旧藏，书画、文玩小品都卖得不错，最后这十柜书作为压轴拍卖，从一开始的二十万起拍，很快就叫到了一百万，然后迅速冲上了两百万。

站在场内左侧的 311 号买家区锦泽坚定不移稳操胜券地一路举到最后一口，最终以 270 万元（含佣金为 310.5 万元）购得。

“先”有王世襄，“后”有明式家具

二十世纪后期，明清家具的艺术价值和商业价值都走向一个高峰，并登上艺术品的殿堂。

这现象产生的原因之一，离不开王世襄先生两本专著的功劳——《明式家具研究》

与《明式家具珍赏》。

在喜爱的事物面前，王世襄天真得像个孩子

从 1945 年开始，王世襄以四十余载坚持不懈、百折不挠的惊人毅力，一点一滴、扎扎实实地积累创造著述所需要的各种条件，终于将《明式家具珍赏》与《明式家具研究》先后出版。

王世襄先生几乎倾半生之力写作的这两部著作，和《明式家具萃珍》《髹饰录解说》《清代匠作则例汇编》《锦灰堆》等著述，做了许多过去没有人做过或做得不够的工作，把明及清前期的家具研究提高到一个令人瞩目的新水平。

王世襄广为人知的明式家具“三部曲”中，《明式家具研究》最早脱稿，有近 30 万字、700 多幅图片。

《明式家具珍赏》则是应香港三联书店之请，从《明式家具研究》中摘录部分内容，并将可以拍到彩色照片的宝物收入图版编著成册而先行出版，随后该书的英、法、德文版先后问世，在台湾也正式出版了中文本，得到中外学术界的广泛重视。北京的一些后起的古家具收藏爱好者，坦言多年来把这部书“翻烂了”。

诚如朱家溍所言：“真正能体现世襄研究成果的是《明式家具研究》。”

该书 1989 年 7 月经由三联书店（香港）有限公司出版，立即受到海内外学术界、收藏界和从业者的高度关注，之后三联书店（北京）又先后四次印刷发行，数量达数万册，仍供不应求。这部“大俗大雅”的巨著，不仅创建了明式家具研究的体系，客观系统地展示了明中期至清前期硬木家具的风采成就，并从历史文化、工艺艺术、审美鉴赏等方面对明式家具的基础研究做了系统的阐述，到现在仍是每位家具业者案头必备的教科书。

王世襄从中国古家具鲜为人知的上世纪 40 年代开始，用 40 年时间完成了明式家具的两部力作，之后又以近 30 年时间促进明式家具文化的传播拓展。

若无王世襄的推动，很难想像中国家具今天会处于怎样的境地。从某种程度上来说，是“先”有王世襄，“后”有明式家具，恐怕也是能成立的。

回到这次拍卖本身来看，将近现代名人生前自用的所有图书一次性打包集中拍卖，这在中国拍卖史上属于第一次，极为罕见，意义非凡，可以视为一次文化事件了。

更难能可贵的是，王老生前自用的这批图书，从书柜到 4000 册图书，是几乎原封不动整柜上拍的。

1996 年，王老整理了珍藏的 80 件家具连卖带捐地送到上海博物馆以后，家里堆积如山的书籍没有地方放了，上海博物馆得知以后，订做了十个书柜送至王老家中。

此后的十几年间，这十个书柜和其内的几千册图书，就成了王老最忠实的伙伴和最温暖的精神寄托，日日相伴，一直到他和夫人先后离世。很多书上都留下了王老亲笔勾画的读书笔记和便签，其间闪烁着他不知多少的思绪火花和情感。

今天来看，王老毕生珍藏散落各处以后，这批生前一直陪伴他每日读书写字的十柜图书，尚能连带书柜一起整体拍卖，并落入同一归处，令人感慨万千。

“单从书本身的价值来看，4000 册图书目前的市场价估摸着在 70 万左右，加上不少书有收藏签名和笔记，散卖价一百多万也很正常。像这样没有拆散，集中在一起连书带柜整体拍卖，意义太特殊了，所以 270 万的落槌价并不贵，还留下了巨大的研究价值。”

311

拍卖结束后，得知买家是来自广东中山的知名家具企业区氏臻品家具新生代掌门人区锦泽先生，我即联系了采访，了解了竞拍前后的故事，这里将采访对话原文照录：

程香：你们在得知这次嘉德拍卖王世襄先生的4000册书籍是什么时候？决心参与竞拍又是什么时候，出于什么样的动力？

区锦泽：得知王老先生的藏书拍卖是大约拍卖的两周前，是从一个圈内的朋友处听说的，当即就决定前来参加拍卖。当然，我们也做了很多准备工作，拿到书目后马上组织了人员进行了价格统计，同时请教了业内的专家老师和朋友，对这批书的心理价位，也是随着认识一步步地提高，可以说我们是很坚定要拿到这批书。决心的原因是我们区氏臻品正在筹备一个古典家具资料馆，筹建资料馆的目的是传承中国传统文化，王老先生作为业界泰斗对我们有很大的影响，刚巧前些天我们也在一个拍卖会上拍了一幅王老先生的字，我觉得这是一种冥冥之中的缘分吧！

程香：竞拍的过程顺利吗？最后促使你们一路咬到底的原因有哪些？

区锦泽：竞拍的过程比我们想象的要顺利很多。这批藏书是最后一件拍品，当时场上的人很多，我们预计会有很多竞争对手。开始拍卖的时候的确有很多人抢拍，但价格加到一百万以后竞争对手只剩下三位，这个时候我们一路坚挺，决心非常强烈，直到最后一刻落槌，当时的成交价比我们心理预期要低，内心很激动。

程香：当最后一刻竞拍成功时，心情和感想如何？

区锦泽：竞拍成功之前心一直提到嗓子眼，这批藏书对我们来说意义很重大，区氏如果有了它，会成为我们资料馆的一笔宝贵财富，所以我们成功的愿望非常迫切。落槌那一刻心才落了地。当时内心最大的感受就觉得这是一件冥冥之中的事，觉得我们和王老真的很有缘分。

程香：去年9月，区氏复刻的一把黄花梨躺椅亮相佳士得上海“当代书房”专场，这把躺椅曾是王世襄先生追寻了四十年的心仪之作，区氏以复刻版参与拍卖，可以看作是对王老的一次致敬。您心中对王老是一种什么样的感情？

区锦泽：是的，王老对我们古典家具从业者来说是心中的一盏明灯。他对古典家具的执着研究和学习是我们学习的榜样。去年在上海佳士得拍卖的那把躺椅是我们对王老的致敬之作，也是我们第一次站在国内最高的平台上亮相。我们区氏对产品品质的追求就是想奉行王老精神的衣钵，把品质做到极致。

程香：听说拍卖后马上有人找你们，想加价购买这批书？你们如何选择的？

区锦泽：是的，确实有好几个人问，实话说我们挺开心的，说明判断正确，大家都认识到这批书的价值了。但是书我们肯定不会再出售了，对我们来说这是最重要的一批收藏，一定要完整地保存下去。这也促使我们开始考虑更多书籍资料的收集，希望对于书籍的珍视和对知识的渴求，成为我们区氏企业文化的重要部分。

程香：今后是否有打算对这 4000 册书籍再做展览或研究？会如何给它们安置一个归宿？

区锦泽：这批藏书对整个区氏来说都意义非凡，它是王老先生一辈子的精神财富，也必将成为区氏和大众的精神财富。本月 13–21 日我们区氏在北京佳士得艺术中心有一个展览，主要是展示一些区氏的作品，很多其实就是依照王老书中的样式复刻的。我们也会从这批书中挑选一部分珍品展出，欢迎大家去现场先睹为快。

佳士得展览过后，这批书将正式收入我们区氏资料馆，作为镇馆之宝，届时我们将把这批藏书对公众开放，让王老的精神普照更多人。在不远的将来，我们还会把这些藏书进行编辑整理，通过科技手段展示给大众，让更多、更广泛的人群有学习的机会。这将是一件非常系统以及艰巨的工作，我们区氏在做好产品的同时会不遗余力地推广中国传统文化，因为热爱古典家具文化，所以希望把我们中华民族最美的文化精神传递出去，这对我们企业来说是一种社会责任。

An Artwork of the Most Profound Sentimental Attachment from Former Collection of Wang Shixiang (Shih-Hsing Wang)

Auction Featured by Wang Shixiang's Collection of Four Thousand Books

By Cheng Xiang

Our social media are awash with all kinds of articles and messages on newly shattered auction records and top-price artworks, sometimes a new record of billions of dollars. With the downturn of domestic stock market and real estate business, the artwork has become an ideal investment option. Therefore, the highlighted featured auctions could draw in the old collectors and also many new faces.

However, aside from all these noise, the only item that touched me the most is a collection of books from Wang Shixiang with a hammer price of 2.7 million RMB (3.105 million RMB with comission).

An Artwork of the Most Profound Sentimental Attachment from former Collection of Wang Shixiang (Shih-Hsing Wang)

This set of books is Wang Shixiang's private collection.

With total 4,000 books (LOT 4689), "Private Book Collection from Wang Shixiang and Yuan Quanyou"appeared in the featured auction of China Guardian.

After Wang Shixiang's demise in the winter of 2009, his lifetime private collections were successively sold in the market, including furniture, ancient Chinese seven-stringed zithers (*Guqin*), Buddhist statues, incense burners, ancient and rare manuscripts and books, bamboo carvings, and pigeon whistles, etc.; whenever the title of "collection of Wang Shixiang"appeared, there would be a sensational bidding spree.

In 2019, after ten-year's circulation in the antique market, Wang Shixiang's private collections were separately collected by different individuals; this set of books is supposedly his last batch of heritage. Compared with the ancient Chinese seven-stringed zither (*Guqin*) of Fuxi's style, named"Melody of the Great Sage"(115 million RMB), and the red sandalwood brush pot carved with patterns of mythical sea creatures (Ming Dynasty, 55.2 million RMB), in the previous auctions, this set of books seemed to be muted in the abustle and astir of this spring auction season and never cause a stir.

The Private Book Collection of Wang Shixiang (4000 volumes) with Original Book Cabinets Appeared in the Spring Auction

At Eight o'clock (at night) of 3rd June, the featured auction "Private Collections from Famous Masters and Literati"of China Guardian kicked off at the presence of numerous collectors and dealers. The latter half of the auction mostly included Wang Shixiang's private collections of paintings, calligraphies, and literati antiques which were sold at ideal prices; the last lot was this set of books which started from 200 thousand RMB, and quickly hit one million RMB, and then hit two million RMB.

The furniture collector, Ou Jinze, holding the No.311 paddle, standing on the left side of the salesroom, kept offering higher price without hesitation, and finally bought the lot with 2.7 million RMB (3.105 million RMB with comission).

Wang Shixiang, the Founder of the Study on Ming-styled Furniture

In the late 20th century, the aesthetic and commercial values of Ming-styled furniture reached its first apotheosis in modern time and the Ming-styled furniture was officially recognized in the art world.

Such achievement was highly dependent on two monographs written by Wang Shixiang, *Connoisseurship of Chinese Furniture: Ming and Early Qing Dynasties* and *Classic Chinese Furniture: Ming and Early Qing Dynasties.*

Passions and Innocence Devoted to the Things Wang Shixiang Loved for Lifetime

Since 1945, with his unremitting persistence and efforts of more than forty years, Wang Shixiang meticulously accumulated all the materials and references required for his monographs on ancient Chinese furniture, bit by bit, and finally published *Connoisseurship of Chinese Furniture: Ming and Early Qing Dynasties* and *Classic Chinese Furniture: Ming and Early Qing Dynasties.*

He had devoted half of the lifetime in these two magnum opus, along with *Masterpieces from the Museum of Classical Chinese Furniture, Notes on the Ancient Record of Lacquerwares, Compilation of Craftsmanship Examples of Qing Dynasty* and *Collection of Miscellaneous Papers*, which have involved many unprecedented researches and studies, brought the study on ancient Chinese furniture (Ming and Qing Dynasties) to a remarkably higher level.

In Wang Shixiang's famous trilogy on Ming-styled furniture, *Connoisseurship of Chinese Furniture: Ming and Early Qing Dynasties* was first finished in draft, with more than 300 thousand characters, and more than 700 illustrations and photos.

And *Classic Chinese Furniture: Ming and Early Qing Dynasties*, an abridged version of *Connoisseurship of Chinese Furniture: Ming and Early Qing Dynasties,* was published first along with many newly taken colourphotos of furniture refered in the book on the request of Hong Kong Sanlian Bookstore; afterwards, it was translated and published in English, French, and German; soon, it was also introduced to Taiwan, and published a traditional Chinese version. Till then, Wang Shixiang's study was widely recognized and received significant attention. Many furniture collectors in Beijing have read this book for so many times that they've lost count.

As Zhu Jiajin once said, "*Connoisseurship of Chinese Furniture: Ming and Early Qing Dynasties,* is the monograph which can best epitomize Wang Shixiang's lifetime study on ancient Chinese furniture."

This book was published in July of 1989, then immediately drew high attention from the domestic and overseas academic community, collectors and relevant practitioners. Then, Beijing Sanlian Bookstore re-printed for four times and sold tens of thousand volumes, yet it was still on short supply. This "elegant and secular"monograph, not only founded the academic system of the study on Ming-styled furniture, but also objectively and systematically illustrated the aesthetic value of the furniture from the middle Ming Dynasty to the early Qing Dynasty, and elaborated a basic study on Ming-styled furniture in terms of history, culture, craftsmanship, and aesthetic appreciation. Till the present day, this book is still the most essential textbook for each practitioner in the furniture business.

From the 1940s when Chinese furniture was known by few people, Wang Shixiang spent forty years to complete his two monographs on Ming-styled furniture, and then took another thirty years to promote the culture of Chinese furniture around the world.

Without his efforts, it is hard to imagine what would become of the study on ancient Chinese furtnire. In a way, Wang Shixiang, is the founder of the study on Ming-styled furniture.

Regarding to this auction, this was the first attempt in Chinese auction history to put a whole set of private book collection from a modern celebrity on one lot. It is rare and also of great significance.

Moreover, the book collection (4000 volumes) were sold with the original book cabinets.

In 1996, Wang Shixiang sold his most preciously collected eighty pieces of ancient Ming-styled furniture to Shanghai Museum at an extremely low price; since then, all the books and references were piled on the ground at home. On learning of the situation, Shanghai Museum ordered ten book cabinets and delievered them to

Wang's home.

In the later years, these ten book cabinets along with several thousand books became Wang Shixiang's most loyal friends and the warmest spiritual consolation, day by day, night by night, until the demise of him and his wife. Many books are marked and commented by Wang Shixiang, which are the true reflection of his thoughts and affections.

Today, after Wang's private antique collections were sold separately in public, this batch of book collection which had accompanied with him for half lifetime being sold wholly with the original book cabinets to one collector is truly a touching moment.

"In terms of the tagged price of these books, these 4,000 books are worthy of 700 thousand RMB, give or take; with signatures and comments and notes, they can be sold separately for more than 1 million RMB; however, to put them in one lot and sell them wholly with the book cabinets, is a meaningful attempt, therefore, the hammer price of 2.7 million RMB is quite reasonable. Moreover, these books also have huge academic value."

After the auction, knowing that the buyer is Ou Jinze, the owner of the famous furniture enterprise in Zhongshan City of Guangdong Province—the Ou's Classic Furniture, I immediately contacted him and made a interview to talk about the stories behind this auction. The interview script is as followed:

Cheng Xiang: When did you learn about that Guardian was going to put Wang Shixiang's 4000-volume book collection on sale? When did you make the decision on bidding for the lot? For what reason?

Ou Jinze: It was about two weeks before the aution, when I was told about the sale of Mr. Wang's book collection by a friend in the furniture business. Then, at once, I decided to bid for the lot. Of course, we've done a lot of works on background checking. We immediately organized a group of people to do the price analysis, meanwhile, we consulted experts and friends about these books to have a better understanding on the price. As I learnt more and more about these books, I became quite determined to buy them, as we were preparing to build an archive for ancient Chinese furniture aming to protect and promote traditional Chinese culture. Wang Shixiang had a great influence in the furniture business. And I just bought a calligraphy work executed by Mr. Wang from an auction the other day. So I believe it was a prophesy and also a pleasant bond.

Cheng Xiang: How did it go? What made you hold out until the end?

Ou Jinze: the bidding was better than we expected. These books made the last lot when there were many people in the salesroom, so we expected more competitors. In the beginning, there were many bidders raising their paddles to push the price up until the price hit one million RMB. Afterwards, there were only three of us bidding for the lot. At this moment, we were even more determined and held out till the end. Knowing that the hammer price was actually lower than my expectation, I was thrilled.

Cheng Xiang: When you finally bought the collection, how did you feel?

Ou Jinze: During the bidding, I was extremely nervous, because these books mean a lot to us; they would be a great fortune for our archive. So we were eager to get this lot. When the hammer struck, I was relieved. I felt that this was destined; we really had a bond with Mr. Wang.

Cheng Xiang: In the last September, your company, the Ou's Classic Furniture, made a replica of a lounge chair made of yellow *Huali (Huanghuali)* wood and brought it to the featured auction "Contemporary Study" of Christie's in Shanghai. This lounge chair was once pursued by Wang Shixiang for forty years. Your company made a repica out of it and brought it to the auction, which can be considered as a homage to Wang Shixiang. How do you feel about Mr. Wang?

Ou Jinze: Yes, Mr. Wang has always been the lantern guiding all the practitioners and collectors of classic Chinese furniture. His persistence and effort on the study and research set an unparalleled example for us. Last year in Shanghai, the yellow *Huali (Huanghuali)* wood lounge chair sold at Christie's is a tribute to Mr. Wang, and also our first appearance on the highest stage in China. We are inspired by his spirit and committed to pursue the highest quality.

Cheng Xiang: I heard that after the auction, there were many people intending to buy these books with higher price, how did you choose?

Ou Jinze: Yes, indeed. Honestly, we are quite happy about it. It says we made the right decision and people started to understand the true value of these books. But we were not going to sell them again. As for us, it is a significant collection; and we intent to keep it in our archive. And it also motivates me to consider to collect more books and references, hoping that the cherishment of books and eager for knowledge can become a significant part of our company culture.

Cheng Xiang: Are you planning to hold an exhibition for these 4000-volume books? Or to do more research on them? Do you have any special arrangement for them?

Ou Jinze: This batch of books are of significant value to the whole company. They are Mr. Wang's lifetime fortune, and still will become a spiritual treasure for us. From 13th to 21st of this month, we are going to hold an exhibition at the art space of Christie's (Beijing), aiming to showcase several works executed by our company. Many of them are replicas of the furniture mentioned or described in Wang's books; also, we will choose part of the important books and documents from this batch of 4000-volume book collection for the exhibition. Hope you can visit us and see for yourself.

After the exhibition, these books will be officially collected in our archive, as the most important treasure for our company. We will open the archive to the public and benefit more people. In the near future, we are going to file and compile them digitally, so that more people could get access to these books. And this will be a quite tough work. We, as a company, will continuously promote our culture and tradition. With passion and love for classic Chinese furniture, we hope that we can pass out the essence of our culture to the world.

王世襄书房

Wang Shixiang Study

永恒的明式家具
MING Furniture in the Forbidden City
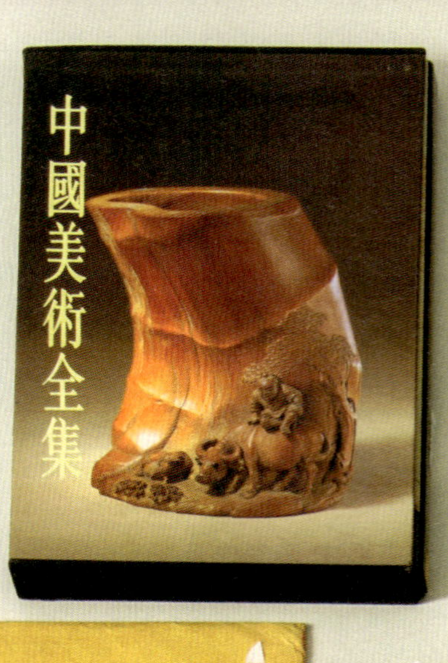
中國美術全集
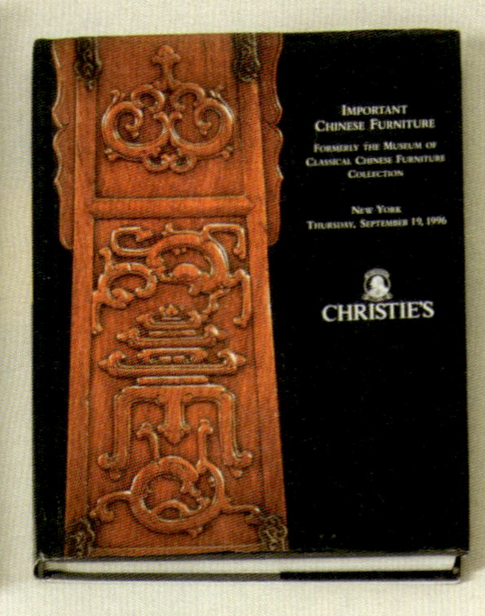
CHRISTIE'S

明式家具珍賞
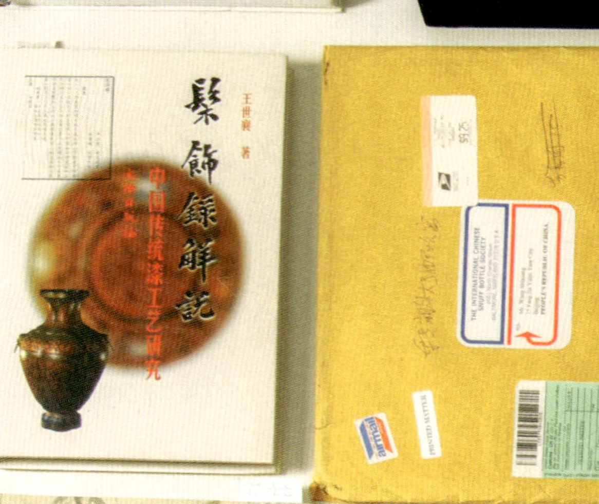
髹飾錄解說
王世襄 著
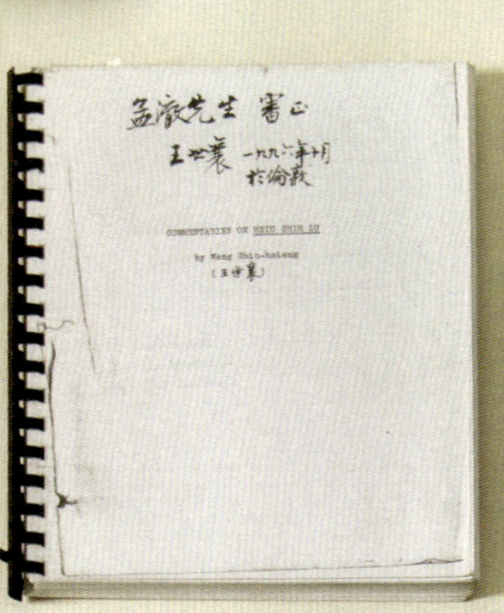

明清傢具集萃
Fine Ming and Qing Furniture in the Shanghai Museum
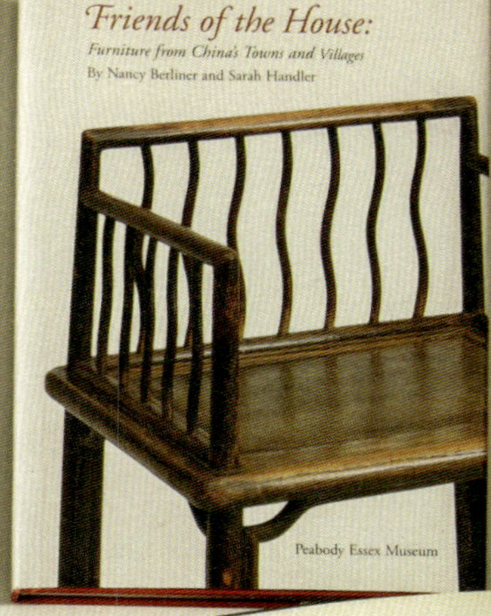
Friends of the House:
Furniture from China's Towns and Villages
By Nancy Berliner and Sarah Handler
Peabody Essex Museum
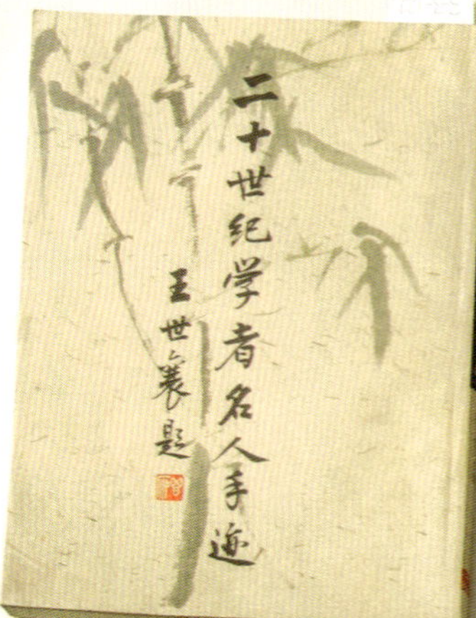
二十世纪学者名人手迹
王世襄题
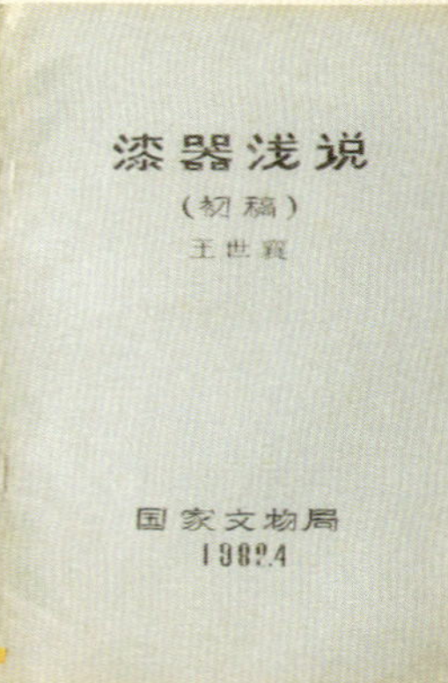
漆器浅说
（初稿）
王世襄
国家文物局
1982.4
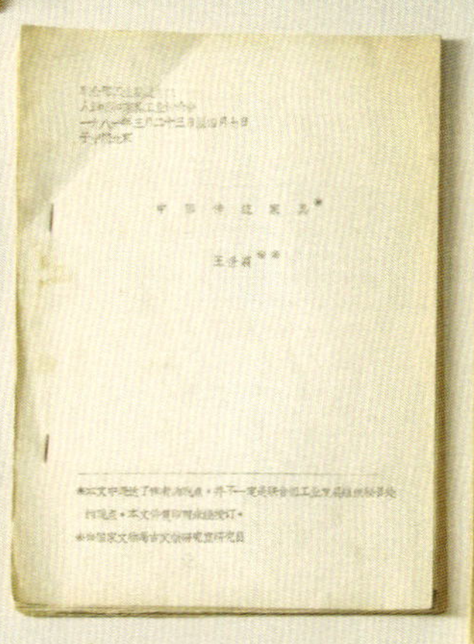
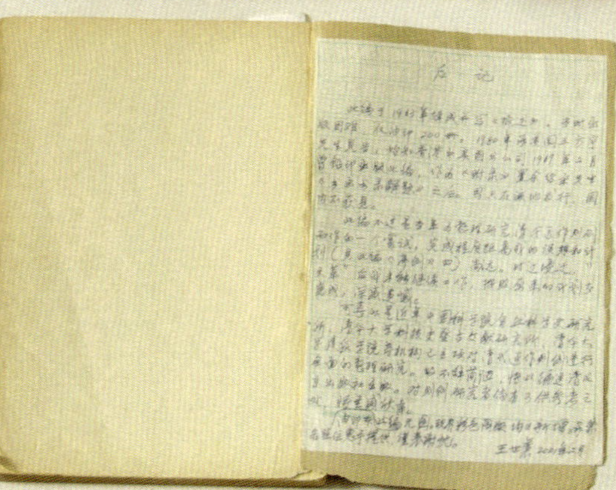

Wang Shixiang
Klassische chinesische Möbel
DVA

Wang Shixiang
Connoisseurship of
Chinese Furniture
Ming and Early Qing Dynasties
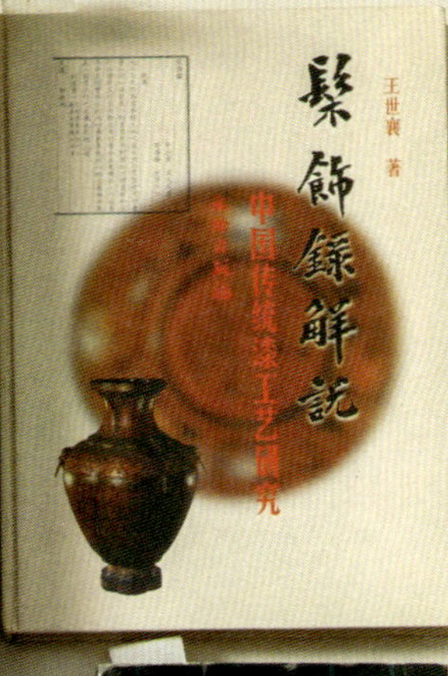
髹飾錄解說
王世襄 著
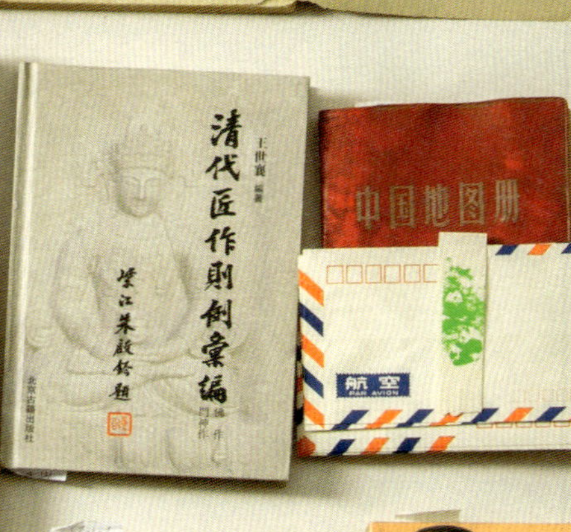
清代匠作則例彙編
中国地图册
航空

王世襄
找一片自己的天地
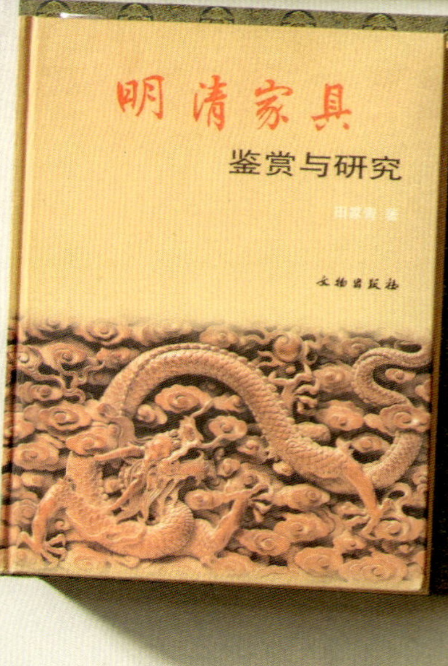
明清家具
鉴赏与研究

中國古典家具博物館圖錄
中國古典家具學會著

中國葫蘆
王世襄 著

竹刻艺术

錦灰堆

竹刻

北京鴿哨

具集珍
髹飾錄解說
Classic Chinese Furniture
Classic Chinese Furniture
Classic Chinese Furniture
Classic Chinese Furniture
中國漆器全集 6 清
中國漆器全集 5 明

竹刻 王世襄編著
明式家具研究 王世襄編著
明式家具萃珍 王世襄編著 袁荃猷繪圖
明式家具研究 王世襄編著
明式家具研究
MOBILIER CHINOIS Editions du Regard
中國鼻煙壺珍賞
The Dr. S. Y. Yip Collection of Classic Chinese Furniture III
Feast by a wine table reclining on a couch
C.L. MA COLLECTION: TRADITIONAL CHINESE FURNITURE FROM THE GREATER SHANXI REGION
Living with Ming - the Lu Ming Shi Collection
明清家具
明清家具

2004

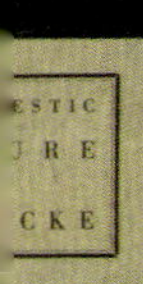

一把让王世襄先生苦苦追寻四十年的躺椅

复刻版亮相佳士得上海

邓彬

醉翁椅属于一种直靠背躺椅式交椅。

早期随游牧民族进入中国的胡床，逐渐改变了中国人原本席地而坐的起居方式，《搜神记》曰："胡床，戎翟之器也。"《风俗通》记载："汉灵帝好胡服，景师作胡床。"

交椅是在胡床基础上的进一步发展，有圆靠背、直靠背两种，直靠背的有些有扶手，有些则无，有扶手的则一般作躺椅式。

明刊本《三才图会》中记述："今之醉翁诸椅，竹木间为之，制各不同，然皆胡床之遗意也。"

《看山阁集·闲笔卷十二·清玩部》也有描述："醉翁椅，斯椅式样颇多，大同小异，置于书屋中，为至美之具，令人相对间，虽欲不醉而不可得也！"

我们今天在古代绘画中时常能看到这种醉翁椅，明代唐寅《桐荫清梦图》（图1）、仇英《桐荫书静图》（图2）、明刊本《三才图会》（图3）中都可见醉翁椅的身影。

上个世纪50年代，南京博物院在苏州东山征集到一具醉翁椅实物，软木髹黑漆，王世襄先生认为和明人所绘形式基本相同（图4）。

中国古代大多数椅子需要正襟危坐，古人强调人的日常行为需要端正身姿，这既是自身修身的要求，也是与众人相处的礼制规范。醉翁椅很少出现在厅堂，无论是《桐荫清梦图》，还是《桐荫书静图》中，画中主人都是独自一人，在树荫或者竹林围绕的书斋中闲坐。

明代陈继儒《太平清话》言："凡焚香、试茶、洗砚……晏坐、翻经……右皆一人独享之乐。"

不同的椅子有不同的坐姿，代表了不同的生活方式。

醉翁椅是躺椅，较之坐椅，人的状态更放松和舒适，是一种介于卧和坐之间的状态。宋代虽然没有出现明代这种交椅式的躺椅，但也有相似功能的椅具出现，南宋刘松年《四景山水图》（图5）中有一白衣高士，其所坐椅子椅面颇深，靠背向后大幅度倾斜；日本大德寺所藏南宋画家周季常和林庭珪的《五百罗汉图》中有一幅表现僧人食瓜的场景，画中有三具扶手躺椅，和《四景山水图》中所绘相似。

王世襄先生在著述《明式家具研究》时，费心搜求实例，"不料过了约二十年"，才在南京博物院库房中看到那件黑漆躺椅，然后又过了约二十年，他在美国加州古典家具博物馆见到一把黄花梨醉翁椅，欣喜万分，并著文介绍。

这把椅子靠背上端用两根立柱分为三格，嵌装绦环板，开扁海棠式透孔，嵌云石，黑质白章，花纹似仙山雾绕、素纸泼墨。后背上方又有荷叶枕托，细藤编制的靠背，舒适逍遥。椅背上端的横材用挖烟袋锅榫与腿足相连，垂扣处向内挑出二小勾，生动有趣，这种做法较为少见，王世襄先生认为这样的处理还"增加了直纹木材的长度，使它不易断裂，对嵌夹绦环板也能起作用"。

图 1 明 唐寅《桐荫清梦图》局部 故宫博物院藏

Plate 1: Ming Dynasty, Tang Yin, *A Daydream under the Shade of Sycamore Tree* (Part), Palace Museum, Beijing

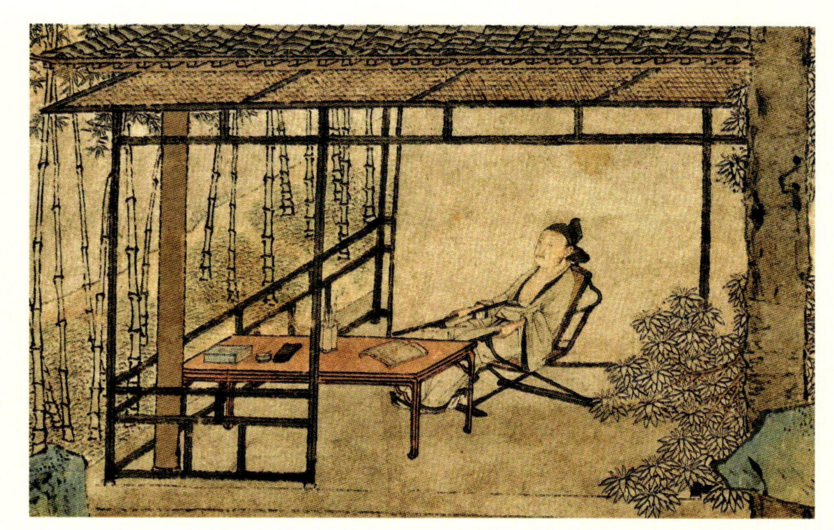

图 2 明 仇英《桐荫书静图》局部 上海博物馆藏

Plate 2: Ming Dynasty, Qiu Ying, *Reading Books under the Shade of Sycamore Tree* (Part), Shanghai Museum

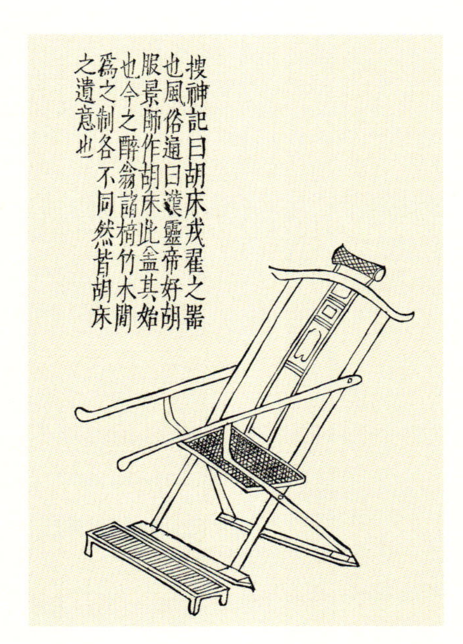

图 3 明 《三才图会》插图中的醉翁椅

Plate 3: Ming Dynasty, Tippler Chair, Illustration from *Illustrated Encyclopaedia of Ming Dynasty*

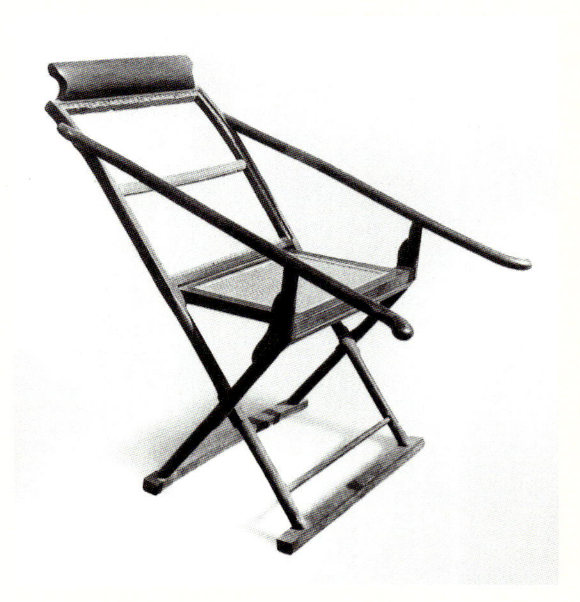

图 4 明 黑漆醉翁椅 南京博物院藏

Plate 4: Ming Dynasty, Black-lacquered Lounge Chair, Nanjing Museum

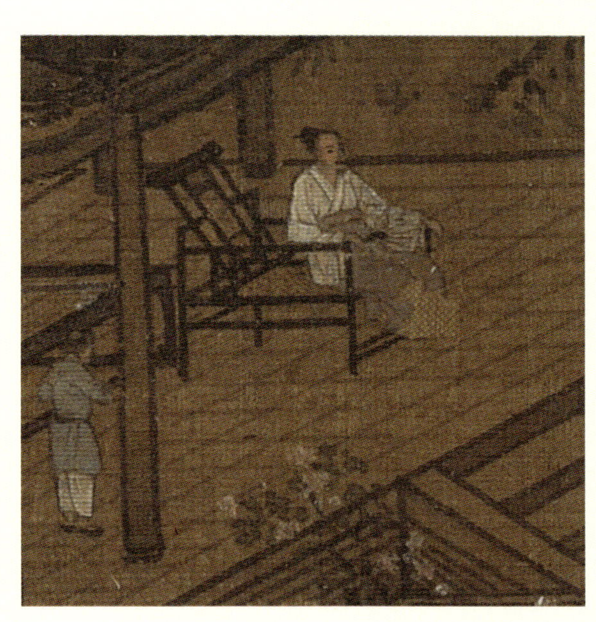

图 5 南宋 刘松年《四景山水图》局部

Plate 5: Southern Song Dynasty, Liu Songnian, *Four Scenes of Landscape* (Part)

古人往往用一个手段能解决好几个问题，这种智慧非多年实践的经验而不可得。

此椅扶手先抱后敞，曲度行云流水，与腿足上截连做托角牙子，既美观又牢固。两足交叉处镶嵌两枚长方形金属构件，中间开圆孔，容纳轴钉。

现有“南粤区氏”数易其稿，反复推敲，用上好的越南黄花梨料，复刻加州古典家具博物馆所藏黄花梨醉翁椅，再次展现古代文人雅具的芳华。

2018 年 9 月 21 日，佳士得上海秋拍的“当代书房”专场 100% 成交，其中，区胜春先生作品“黄花梨躺椅”，经过场内外激烈竞投后，最终以 504000 元落槌成交（含佣金）。

佳士得亚洲区总裁魏蔚表示：“本季成绩让我们倍感振奋，知名和新锐艺术家得到了中国和国际藏家的认可，展示出市场对于优秀艺术作品的兴趣。”

区氏躺椅的成功拍卖，对当下处于凛冬中的古典家具市场，无异于是吹来了一缕春风，提振市场，鼓舞人心。无论市场如何变幻，致力于传统与当代工艺深耕的品牌和作品，一定会赢得市场的认可。

至此，区氏透过这把穿越古今的中国躺椅，一则逆势而上，表达南粤企业对市场的信心；二则以 1:1 复刻经典，完成向王世襄先生的致敬。历史上曾被文人推崇备至的一款独特椅具，放置今天的书房，真可谓：“为至美之具，令人相对间，虽欲不醉而不可得也！”

黄花梨躺椅
110 x 65 x 100 cm
RMB: 504,000
FIRST OPEN | Shanghai
佳士得拍卖（上海）有限公司
2018 年 9 月 21 日

Yellow *Huali (Huanghuali)* Wood Lounge Chair
FIRST OPEN | Shanghai
Christie's (Shanghai)
21st September, 2018

A Lounge Chair Pursued by Wang Shixiang (Shih-Hsing Wang) for Forty Years

The Exhibition of the Replica at Christie's in Shanghai

By Deng Bin

The tippler chair is a kind of folding chair with straight backrest.

In early time, the folding stool brought into China by nomads had gradually changed Chinese people's ways of sitting or resting on the ground level. According to *In Search of the Supernatural*, "the folding stool, a furniture from the west"; and in the *History of Customs*, "as Emperor Ling quite likes wearing clothes of ethnic minorities from the west, Jing Shi made him a folding stool."

The folding chair is developed on the basis of folding stool, with round horseshoe back or straight back; some straight-back folding chairs are installed with armrests, which are usually built in lounge chair style.

According to *Illustrated Encyclopaedia of Ming Dynasty* (published in the Ming Dynasty), it says that, "today's tippler chairs, made of bamboo and wood, with various forms, actually are developed from the folding stool." In the *Collection of Mountain-seeing Pavillion*, "the tippler chairs, with various types and forms, are all developed from a same structure; they can be placed in the study as an ideal sitting furniture; in the presence of such furniture, one can only indulge in its beauty!"

Such tippler chairs can also be found in those ancient paintings, like *A Daydream under the Shade of Sycamore Tree* by Tang Yin (Plate 1), *Reading Books under the Shade of Sycamore Tree* by Qiu Ying (Plate 2), or illustration from *Illustrated Encyclopaedia of Ming Dynasty* (published in the Ming Dynasty) (Plate 3).

After 1950s, Nanjing Museum acquired an ancient tippler chair in Suzhou City, which is made of softwood coated by black lacquer; Wang Shixiang indicated that the chair is built in the same structure with those of the chairs painted by the people in the Ming Dynasty (Plate 4).

In ancient China, most chairs are built in a structure to force people to sit properly. They mostly emphasized the upright postures as a reflection of one's manners and education, and also the ritual regulation during social activities. So the tippler chair is rarely seen in the living room; either in the *A Daydream under the Shade of Sycamore Tree* by Tang Yin (Plate 1), or in the *Reading Books under the Shade of Sycamore Tree* by Qiu Ying, there is only one figure in the painting, implying a private scene.

In the *Notes on Daily Life* written by Chen Jiru in the Ming Dynasty, it says that, "sitting on a tippler chair, all by myself, I can fully enjoy the life and do whatever I want, like burning incense, having a cup of tea, washing my inkstone, taking a nap, reading the Buddhist scripture."

Different chairs require different sitting postures, which represent different life styles.

The tippler chair is a kind of lounge chair, so it is more comfortable than other traditional chairs; people tend to be more relaxed while sitting (a state between sitting and lying) in such chair. In the Song Dynasty, although the tippler chair hadn't appear, there was a similar chair with the same structure. According to the painting *Four Scenes of Landscape* by Liu Songnian (Plate 5), there is a scholar wearing white robe, sitting on a chair with larger and deeper seat panel, and big slope in the backrest; and in the painting *Five Hundred Arhats* by Zhou Jichang and Lin Tinggui (collected in Daitokuji Temple, Japan), there is a panel depicting an arhat eating a slice of watermelon, with three lounge chairs (with armrests) placed nearby which are similar to the chair in the painting *Four Scenes of Landscape*.

While writing the *Connoisseurship of Chinese Furniture: Ming and Early Qing Dynasties*, Wang Shixiang searched around for an actual piece of similar lounge chair. After about twenty years, he happened to bump into that black-lacquered lounge chair in the storage of Nanjing Museum; in another twenty years, he finally met a yellow *Huali (Huanghuali)* wood lounge chair (tippler chair) in Museum of Classical Chinese Furniture in California, USA, he was thrilled and wrote an article about it.

The upper part of the backrest of this lounge chair are divided into three compartments by two short struts and embedded with decorative panels carved with begonia-shaped tracery which are inlaid with marble slabs, with black-and-white natural veinings, resembling the fairy mountain looming in the cloud, or a ink and wash landscape painting. Above the backrest installs a small lotus leaf-shaped headrest; the soft backrest cushion made of woven rattan can provide a rare and extreme comfort. The top horizontal rail of the backrest is connected to the legs with pipe-style joints; the jammed buckles are curved inward as two small hooks, which is a structure rarely seen among others. Wang Shixiang assumed that such structure can "increase the length of the components so as to prevent them from breaking or cracking, and also help to fix the decorative panels".

The ancient people can usually solve several problems through one approach; such wisdom definitely requires experience of many years' practice.

The two armrests form a hugging posture from the backrest and then splayed outward at ends, drawing an elegant silhouette in the air and connecting to the upper parts of two rear legs by corner plates which function as structural components and also decorations. Two rectangular metal components are embedded at the crossing points of two legs; they are bored a hole to contain the pivot nails.

Today, the Ou's Classic Furniture from southern Guangdong Province, through countless modifications and researches, applied the top Vietnamese yellow *Huali (Huanghuali)* wood to the replication of this yellow *Huali (Huanghuali)* wood lounge chair (tippler chair) from Museum of Classical Chinese Furniture in California, USA, in hope of representing the true elegance of the ancient Chinese furniture.

On 21st September of 2018, the featured auction "Contemporary Study" of Christie's (Shanghai) achieved a white glove result; Among them, Ou Shengchun's work— yellow *Huali (Huanghuali)* wood lounge chair, was sold at 504 thousand RMB (with commision).

Rebecca Wei, the president of asia (Christie's), expressed her excitement, "the brilliant results achieved this season are really inspiring; many famous and new artists received extensive recognition, which shows the interests of the market toward to excellent works."

The success deal of the lounge chair made by Ou's Classic Furniture, brings a wisp of spring breeze to the winter of classic furniture business with much of inspirations and encouragement. Despite the ever-changing market, as long as devoting to the traditional craftsmanship and also the development of contemporary technologies, one (individual or company) shall finally conquer the market.

So far, through this classic Chinese lounge chair, Ou's Classic Furniture reversed the bad situation and expressed their confidence to the market; secondly, they made a replica of a piece of classic furniture with meticulous details and in full size to pay tribute to Wang Shixiang. An unique chair once revered by many literati and scholars through history, now is placed in today's study, which truly is, "in the presence of such furniture, one can only indulge in its beauty!"

橱与柜的名称由来与形制嬗变

邓彬

2019 年 9 月 20 日，上海佳士得拍卖预展上有一对新制铁力木圆角柜，尺寸小巧可爱，材美工妍。

门芯板为瘿木，柜顶转角为圆，柜帽喷出，侧角显著，这是典型的明式圆角柜的样式。

关于圆角柜有多种不同的名称，北方地区大多称其为“面条柜”，苏州地区则称其为“大小头橱”，橱和柜是通用的，但在中国历史上，早期的橱和柜是有分别的。

储物类的家具统称为庋具，庋是收纳的意思，庋具是储物的家具。宋人戴侗《六书故》中说：“今通以藏器之大为柜，次为匣，小为椟。”庋具中尺寸大、储物多的家具即是柜子。

柜是简化字，原本是“匮”加木字偏旁，“匮”的本意是储存贵重物品的容器，山东沂南汉墓画像石就出现了柜的图像（图 1），汉代的柜一般四面垂直，无帽顶，四足直接落地。

汉代橱和柜有区别，北宋沈括在《梦溪笔谈・补笔谈卷二》中谈到：“阁者，板格。以庋膳馐者，正是今之立馈。今吴人谓立馈为厨者，原起于此，以其贮食物也，故谓之‘厨’。”

“厨”通“橱”，早期的橱是在厨房里贮放食物的家具。

汉代的橱有屋形橱顶，一般有两扇可以开启的门，辽宁辽阳棒台子屯东汉墓壁画中描绘有汉代橱的图像（图 2），北宋《大宋重修广韵》中描述，“橱，帐也，似厨形也。”

橱形似室内张挂的帷帐，如屋宇，早期的橱是厨房中贮存食物的家具。

两晋时期出现了书橱，原本只用作储存食物用的橱有了新的用途，晋《东宫旧事》记载：“皇太子初祥，有柏书橱一，梓书橱一。”

图 1　山东沂南汉墓画像石出现柜的图像

Plate 1: Cabinet Image Appeared on Stone Relief of Han Tomb in Yinan County of Shandong Province

图 2　辽宁辽阳棒台子屯东汉墓壁画中的汉代橱

Plate 2: Cupboard of Han Dynasty Appeared in Murals of a Tomb of Han Dynasty at Liaoyang City of Liaoning Province

到了唐代，柜和橱依然不同，日本正仓院则保留有圣武天皇生前的遗物，主要是他的生活用具，在他去世后，皇后将他的遗物捐赠给了东大寺，捐赠时入库的献物账现在还保存着，所以我们可以得知这些物品当时的名称。其中有一件赤欟木御厨子（图3），有帽顶。而另一件玉虫厨子，其形正是如屋宇一般。正仓院藏有一件赤漆密陀绘云兔柜子（图4），柜体四平，四足设在柜体外侧。

1955年，西安出土了一件唐代的三彩瓷柜（图5），虽然是明器，但系忠实于实用器而制作，柜体是四平样式，没有柜帽。所以，唐代的柜和橱仍然属于两种不同形制的储物家具。

初唐李延寿在《南史·陆澄传》写道："澄当世称为硕学，读《易》三年不解文义，欲撰宋书竟不成。王俭戏之曰：陆公，书厨也。"讽刺只会读书不会应用的人是两脚书橱。

唐代白居易有一首诗写道，"破柏作书柜，柜牢柏复坚。收贮谁家集，题云白乐天"。可见唐代的橱与柜都出现了专事贮存书籍的用途。

从前文引述沈括的文字我们得知，宋人将"立柜"称为"厨"，之所以在柜前加一个立字，是因为之前的柜是矮型家具，宽度要大于高度，而宋代出现的一类所谓的立柜显然是改变了这样的状态，柜是"立"起来的，和橱一样。

宋代的绘画上所见大多是橱，多为盝顶，柜顶削出四个倾斜的斜面，有两扇开启的小门，盝顶亦是对建筑屋顶或者帷帐的模仿，体现礼制观念，汉唐时期大多用作于小型箱匣。北宋出现了较大的盝顶家具，有两扇小门开启，显然已经属于橱的性质。

图3　赤欟木御橱　日本正仓院藏
Plate 3: Red Lacquered Wooden Imperial Cabinet, Shōsō-in, Japan

图4　赤漆密陀绘云兔柜　日本正仓院藏
Plate 4: Red Lacquered Wooden Cabinet Carved with Bas-Relief Patterns, Shōsō-in, Japan

图5　唐三彩瓷柜
Plate 5: Tri-colour Glazed Porcelain Cabinet Unearthed In Xi'An City

图 6　北宋《文会图》中的橱
Plate 6: *Gathering of Literati* executed by Zhao Ji, the Emperor Huizong, Northern Song Dynasty

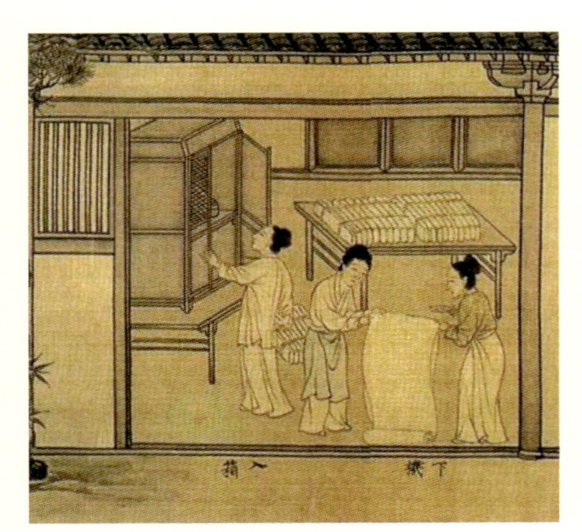

图 7　南宋《蚕织图》中的橱
Plate 7: *Silk-making* by anomymity, Southern Song Dynasty

图 8　南宋　刘松年《唐五学士图》
Plate 8: *Five Scholars of the Tang Dynasty* executed by Liu Songnian, Southern Song Dynasty

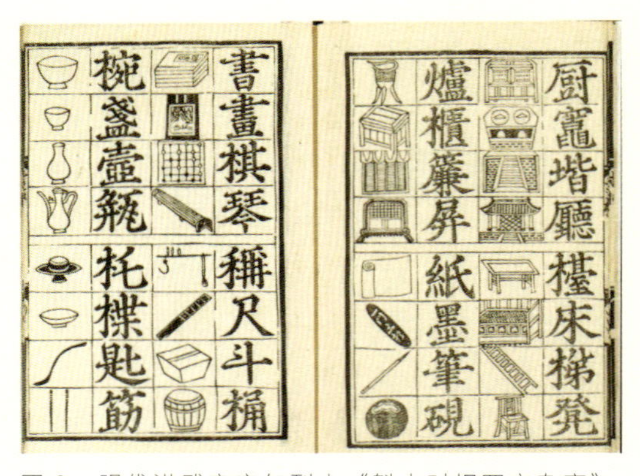

图 9　明代洪武辛亥年刊本《魁本对相四言杂字》
Plate 9: Woodcut Illustrations from *Illustrated Dictionary* published in 1371 (Hongwu Period of Ming Dynasty)

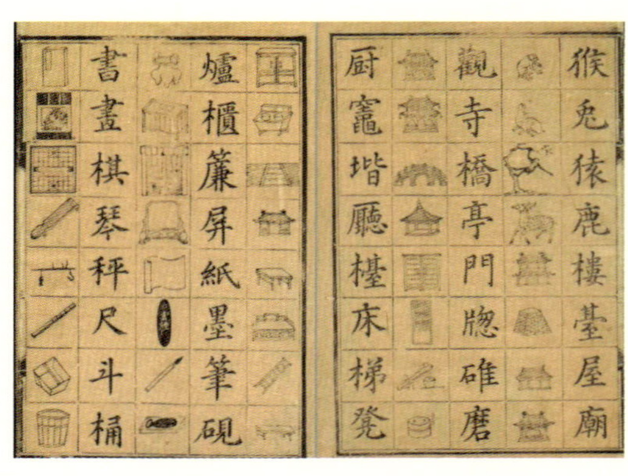

图 10　《新编对相四言》，约为 1506 年后刊本
Plate 10: Woodcut Illustrations from *Verses on Character Learning* published after 1506

北宋宋徽宗《文会图》(图 6)、南宋佚名《蚕织图》(图 7)、刘松年《唐五学士图》(图 8) 都描绘有这样的橱。

宋代还出现了书橱中设抽屉的做法,宋周密《癸辛杂识》记载:“昔李仁甫为《长编》,作木厨十枚，每厨作抽替匣二十枚，每替以甲子志之。”在橱中设抽屉，是宋代橱柜家具制作的新突破。

元代沿用宋制，甚至在明初期依然没有改变。明代洪武辛亥年刊本《魁本对相四言杂字》，用木刻版画分别表现了橱与柜（图 9），柜的形象和汉唐时期近似，而厨的样子和宋代是相似的，盝顶，设两扇小门。明代初年下葬的朱檀墓中未出现橱柜，只有放置在台案上的衣箱。汉唐至明初期，衣物一般放在箱、箧、笥、笈等庋具中，尚未出现存放衣物用的橱柜。

橱柜家具的大变革是在明中期以后。

由于黄花梨等硬木材料的大量使用，木工工具的进一步完善，经济和社会诸多因素的作用之下，晚明出现了许多新的家具样式，达到了中国古代家具制作新的高峰。

美国哥伦比亚大学图书馆所藏明代蒙学图书《新编对相四言》，大致是 1506 年后刊本（图 10），所列的橱是一只四抹门圆角柜的样子，这是目前已知最早的明式圆角柜图像，而柜的形态依然延续汉唐时期的样子。

仇英创作于 1542 年至 1545 年的《清明上河图》（图 11，图 12）描绘的是明代苏州的景象，画中多次出现四抹门圆角柜（橱）。而成书于 1607 年的《三才图会》中所绘制的橱样（图 13），有四门和两门两种圆角柜，都是四抹门式。

明晚期还出现了圆角柜柜门用独板的例子，卒于 1589 年的潘允徵，其墓葬中出土

图 11　明　仇英《清明上河图》中的圆角柜
Plate 11: Round-cornered Cabinet in *Riverside Scene at Tomb-Sweeping Festival* , executed by Qiu Ying, Ming Dynasty

图 12　明　仇英《清明上河图》中的圆角柜
Plate 12: Round-cornered Cabinet in *Riverside Scene at Tomb-Sweeping Festival*, executed by Qiu Ying, Ming Dynasty

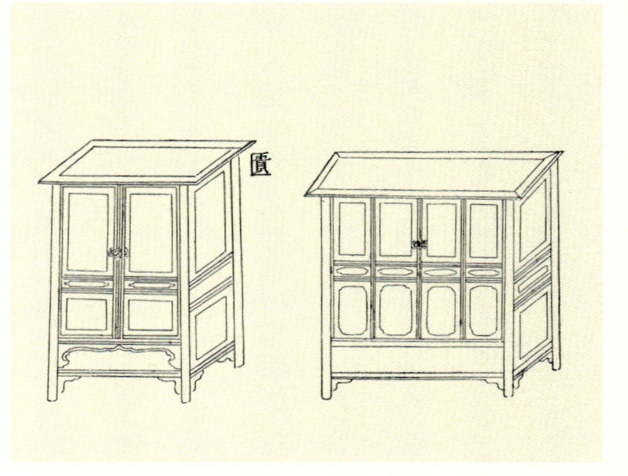

图 13　明　《三才图会 · 器用十二卷》中所绘制的橱样
Plate 13: Cupboard Image from *Illustrated Encyclopaedia of Ming Dynasty* (published in 1607)

的一对圆角柜明器，其柜门即是独板（图 14）。明式方角柜有明代实物传世，北京故宫藏有一对万历年款的方角药柜。

橱至少在明正德年以后，出现了明式圆角柜的形制，有柜帽，柜顶转角为圆，侧角明显，设两扇柜门，门四抹，这种样式的圆角柜一直持续到清早期。上海潘允徵墓出土的柜门独板式的圆角柜是四抹门圆角柜的演进样式，出现的时代不晚于万历年间。柜至少在明嘉靖年以前，大多一直延续早期柜子的形制，明晚期开始出现明式方角柜。

值得注意的是，《三才图会》所绘制的圆角柜，书中刊为“匮”，匮通柜，而差不多同时期成书于万历年间的增编版《鲁班经匠家镜》所列“衣橱样式”，其造法也是圆角柜式，所以橱与柜的名称在此时已经混用，没有区别。但早期橱柜所指向的不同样式依然被延续下来了。

明式柜类从构造方式而言分两大类，一种是圆角柜，有柜帽，其更多是延续早期橱类家具的做法；另一种是方角柜，无柜帽，更多的是继承早期柜类家具的特征。

图 14　明　潘允徵墓葬中出土的圆角柜明器
Plate 14: Burial Object Of Round-Cornered Cabinets, unearthed from the tomb of Pan Yunzheng (died in 1589)

参考书目

王世襄：《明式家具研究》，生活 · 读书 · 新知三联书店。
陈增弼：《传薪——中国古代家具研究》，故宫出版社。
孙机：《中国古代物质文化》，中华书局。
王正书：《明清家具鉴定》，上海书店出版社。

The Name Origin and Form's Evolution of Cupboard and Cabinet

By Deng Bin

On 20th September of 2019, in the preview of Christie's (Shanghai), there were a pair of newly made lignumvitae round-cornered cabinet, built in adorable medium size and elegant structure, and made of refined material.

Door panels are made of burl wood; the top panel is built with rounded protruding corners, which is a typical round-cornered cabinet.

It has many different names across China: in the north, it is called "noodle cabinet"; in the area of Suzhou City, it is called "transition cupboard"; the words cabinet and cupboard are now convertible, yet in ancient China, early cupboard and cabinet did have some differences.

The furniture which is used to store things is called storing furniture, meaning to collect or to put away. According to *Analyses on Chinese Characters* written by Dai Tong in the Song Dynasty (960-1279), it says that, "today, the large storing furniture is called cabinet; the medium sized is called case; and the small sized is called casket."So the large storing furniture is cabinet.

Cabinet (柜), is a simplified Chinese character; the original traditional character is "匱"with a "木"on the left, as"櫃". The original meaning of "匱"indicates a utensil used to store up valuables, according to the image of early cabinet found in stone relief of Han Tomb in Yinan County of Shandong Province (Plate 1): cabinets in Han Dynasty are mostly built with straight panels without cabinet head or legs and feet.

In the Han Dynasty (206 B.C.-220 A.D.), there were differences between the cupboard and the cabinet, according to *Dream Pool Essays* written by Shen Kuo in the Northern Song Dynasty, "cupboard, or shelved cupboard, was used to store food, which is today's standing cabinet; therefore, today's people of Wu area call the standing cabinet, cupboard; because it is used to store food, so is called cupboard."

"厨" and "橱"were convertible; the early cupboard was specifically used to store food in the kitchen.

The cupboard in the Han Dynasty was built with roof-styled top and two doors, according to the image of cupboard in the Han Dynasty depicted on murals of a tomb of Han Dynasty at Liaoyang City of Liaoning Province (Plate 2), and *Rhymes and Tones of the Song Dynasty* written in the Northern Song Dynasty, "the cupboard, looks like a tent, with a cabinet structure."

The early cupboard quite resembled a tent put up in the room, or a house; they were used to store the food in the kitchen.

During the Western and Eastern Jin Dynasty (265-420), the book cabinet appeared with a new function aside from food storing in the kitchen, according to *Memories of East Palace* written in the Jin Dynasty, "the crown prince made the drafts of one cypress bookcase and one catalpa bookcase."

In the Tang Dynasty (618-907), the cabinet and the cupboard were still quite different from each other. Today, Shōsō-in still preserves all the items that Emperor Shomu once used or owned, mostly are daily supplies. After his demise, the empress donated all the items and heritage to Todai-ji Temple, along with the tribute list, therefore we can get the names of those donated items. Among them, there are a red lacquered wooden imperial cabinet (Plate 3), with a hat-styled top, and another cabinet built in house-style. And in the middle storage of Shōsō-in, there is a red lacquered wooden cabinet carved with bas-relief patterns (Plate 4), with flat panels and four legs attached to the cabinet body on two sides.

In 1955, there was a tri-colour glazed porcelain cabinet unearthed in Xi'an City (Plate 5), although it was a burial object, this porcelain cabinet was modeled from the actual furniture, with flat-corner structure (corners butted with mitered corner bridle joints, also called cubic-cone rice-pudding joint), without the cabinet hat. Therefore, in the Tang Dynasty, the cabinet and the cupboard still were storing furniture of different structures.

According to *History of Southern Dynasties* written by Li Yanshou in the early Tang Dynasty, "Lu Cheng is recognized as a knowledgeable scholar, however, he read *The Book of Changes* for three years and still didn't understand the meaning, and couldn't write *History of the Song Dynasty*. Wang Jian laughed at him and said, 'Mr. Lu is just a book cupboard.' (a satire on Lu Cheng, implying that he can just read books but can not apply knowledge to practice.)"

Bai Juyi, the famous poet in the Tang Dynasty, once wrote that, "the book cabinet made of cypress, is as strong as the hard cypress, to collect books in the cabinet, which are written by Bai Juyi."Clearly, in the Tang Dynasty, the cupboard and the cabinet were all used to store books.

According to previous quote from Shen Kuo, in the Song Dynasty, people called the "standing cabinet" as "cupboard". The character "立"(standing) added in front of cabinet implies that the previous cabinet was short furniture, with its width greater than the height; however, the standing cabinet appeared in the Song Dynasty obviously changed the structure: the original short cabinet stood up, like the cupboard.

Many cupboards can be seen in the paintings of Song Dynasty, mostly with truncated pyramid head and four slopes on the top, and two doors in the front. The truncated pyramid top is also modeled from the house roof or tent to reflect the hierarchy concept, and mostly appied in the Han and Tang Dynasties on small-sized caskets; in the northern Song Dynasty, there appeared larger furniture with truncated pyramid top and two small doors, which was clearly features of the cupboard.

Similar cupboards also can be seen in the paintings of *Gathering of Literati* (Plate 6) executed by Zhao Ji, the Emperor Huizong, *Silk-making* (Plate 7) by anomymity in the Southern Song Dynsty, and *Five Scholars of the Tang Dynasty* (Plate 8)executed by Liu Songnian.

In the Song Dynasty, some book cupboards were installed with drawers inside, according to *Essays and Other Articles* written by Zhou Mi in the Song Dynasty, it says, "Li Renfu once ordered ten wooden book cupboards for storing his series *Complemented History As A Mirror*, with each cupboard installed twenty drawers marked by numbers." To set drawers inside the cupboard is an original creation in the Song Dynasty.

In the Yuan Dynasty, people followed the previous structure, even till the early Ming Dynasty. According to *Illustrated Dictionary* published in 1371 (Hongwu Period of Ming Dynasty), the woodcut illustrations respectively depict the images of cupboard and cabinet (Plate 9): the cabinet is similar to the ones made in the Han and Tang Dynasties; and the cupboard is more like those ones made in the Song Dynasty with truncated pyramid top and two doors. Moreover, in the tomb of Zhu Tan who was buried in the early Ming Dynasty, there is no signs of cupboard or cabinet, only having some clothes placed on the table. From the Han and Tang Dynasties to the early Ming Dynasty, clothes were commonly stored in the chest, casket, bamboo-plaited suitcase, or box; there weren't any cupboards or cabinets used to store clothes.

The great transformation of cupboard and cabinet happened after the mid Ming Dynasty.

With the extensive appliance of many hardwood material like yellow *Huali (Huanghuali)* wood, and the further development of tools and craftsmanship, along with other economic or social changes, in the late Ming Dynasty, China experienced a new apotheosis of classic furniture making; and during that period of time, many new types of furniture appeared.

According to the abc book for kids in the Ming Dynasty, *Verses on Character Learning* (collected in the library of Columbia University) published after 1506 (Plate

10), the cupboard depicted is similar to a round-cornered cabinet; so far, this is the earliest-known image of the Ming-styled round-cornered cabinet; and the cabinet still follows the style in the Han and Tang Dynasties.

According to the painting *Riverside Scene at Tomb-Sweeping Festival*, executed by Qiu Ying between 1542 to 1545, which depicts the urban-rural landscape of Suzhou City in the Ming Dynasty, the round-cornered cabinet (cupboard) with four door panels appears in the painting for many times; and according to *Illustrated Encyclopaedia of Ming Dynasty* (published in 1607), the cupboard was built either with four door panels or two door panels.

In the late Ming Dynasty, there appeared the round-cornered cabinet with door panels made out of one-piece material. In the tomb of Pan Yunzheng (died in 1589), a pair of burial objects of round-cornered cabinets were installed with door panels made out of one-piece material (Plate 14). There are a pair of existing Ming-styled square-cornered medicine cabinets (with the seal of Wanli Period) collected in Palace Museum (Beijing).

After the Zhengde Period of Ming Dynasty, there appeared the Ming-styled round-cornered cupboard, with cupboard hat, round-cornered top board, two door panels with four compartments; such style lasted till the early Qing Dynasty. The pair of burial objects of round-cornered cabinets were installed with door panels made out of one-piece material unearthed from the tomb of Pan Yunzheng are evolved version of basic round-cornered cabinet, which appeared no later than the Wanli Period. At least before the Jiajing Period of the Ming Dynasty, the cabinet had been following the early structure; then in the late Ming Dynasty, Ming-styled square-cornered cabinet appeared.

It's worth noting that, the round-cornered cabinet drawn in the book *Illustrated Encyclopaedia of Ming Dynasty* was labeled "匮", convertible to "柜"; yet in the book *Lu Ban's Notes on Carpentry* published in the same Wanli Period, the cupboard was also depicted as the round-cornered cabinet. So the names of cupboard and cabinet were convertible during this time, with no difference, yet the early differences between the two types of furniture are followed.

In terms of structure, the Ming-styled cabinet can be divided into two categories: one is round-cornered cabinet (with cabinet hat), which mostly follows the structure of early cupboards; the other is square-cornered cabinet (without cabinet hat), which mostly follows the structure of early cabinets.

REFERENCE:

Wang Shixiang: *Connoisseurship of Chinese Furniture: Ming and Early Qing Dynasties*, Sanlian Publishing House

Chen Zengbi: *Heritage—the Study on Ancient Chinese Furniture*, Palace Museum Press

Sun Ji: *Tangible Cultural Heritage of Ancient China*, Zhonghua Book Press

Wang Zhengshu: *Authentication of Furniture in Ming and Qing Dynasties*, Shanghai Bookstore Publishing House

铁梨木嵌瘿木圆角柜（一对）

69×42×99 cm

RMB: 114,000

开创 | 上海：随艺而居

佳士得拍卖（上海）有限公司

2019 年 9 月 21 日

Lignumvitae Round-cornered Cabinet Embedded with Burl Panels (A Pair)
OPENING | Shanghai: Live With Art
Christie's (Shanghai)
21st September, 2019

佳士得在北京打造了一个理想中的『家』

程香

视线左侧置一对黄花梨圈椅，中央有柔软舒适的沙发可供围坐。面前安置了一几一凳一拳石，还有两只宫灯，靠墙立着一对亮格柜，凑近细看门板系一木而做，质朴辽阔，低调奢华。

茶室中央是一张四面平式宽大茶桌配四出头官帽椅，造型简练利落。茶桌右侧角落有一带底座圆角柜和一只三弯腿五足圆香几，灵动自然。茶桌前摆放的是丹麦设计师汉斯韦格纳1944年设计的“中国椅”，其灵感来自明式圈椅，充分借鉴了圈椅的舒适性和线条感。茶桌上摆着紫砂壶、茶杯、建水、盖置、茶匙、茶盘等各式茶道具。

对面墙边摆着两张柯步西耶设计的LC-2黑色沙发，用不锈钢和真皮制成，刚硬和柔软和谐共存。沙发侧面的墙上悬挂的是一幅摄影师陈漫大学时期的作品《双米奇2号》。

中西方气息在空间内无声交融，可以坐语道德，畅怀舒啸……

关于中式家具在当代语境下的呈现方式，关于古为今用的诠释角度，关于艺术生活化的有趣探索，关于中西文化的交融和碰撞，还有什么可以在当下想象和发生?

今年6月，佳士得在北京搭建了五个实景空间，营造了一个理想中的“家”。

一　Old is New

2015年，佳士得在纽约还原了安思远的华美大宅，将一代中国古董大亨的生前家居空间，呈现在世人面前，也将西方收藏大家的生活起居文化带回中国。

安思远之后，佳士得又陆续将一众西方重要藏家的家居空间呈现在拍前预展，掀起一波又一波的时尚收藏风潮。

而在中国，随着接收到的收藏信息日趋丰富，国人对艺术日用和陈设美学方面，也有了更高的标准和更深的渴求。

2019年6月13日－21日，佳士得以私人洽购的方式，在北京佳士得艺术中心模拟了一个精致美好的“家”。

这个“家”，由客厅、偏厅、茶室、佛堂、书房组成，以古为今用的方式，围绕生活、艺术、设计三个层面，展开一系列有益的探讨和尝试。

此次“汲古涵今——当代书房艺术”展，精选了全球90余件艺术品与生活器具，围绕书房文化与书房功能的延伸展开。

展览从生活切入，将区氏制作的中国明式经典家具、包豪斯思潮影响下的现代主义家具、北欧设计、意大利当代设计的家具和家饰品，以及吴昌硕、吴湖帆、王原祁、金俊明等古代文人画家作品，和井上有一、杉本博司等日本艺术家作品，自然有机地陈列在一起，营造出一个适用于生活又具有高度审美的艺术化生活场景，移步换景，值得品味。

二　再现“王世襄书房”

大家一定还记得近期热文《王世襄最有温度的旧藏：生前自用的 4000 册图书整体拍卖》，王世襄生前自用的十柜图书 4000 册，6 月初刚刚由中国嘉德拍卖，而买方正是参与此次“汲古涵今”展览的区氏家具。

此次展览的书房空间，陈列了数十本王世襄先生藏书（部分），成为一大亮点。

画案上的有些书摊开了，夹着某封书信或邀请卡或报纸剪报，王老使用的痕迹清晰可见。静静凝视间，仿佛能见到老先生挑灯夜下执笔静读的身影，不禁令人肃然起敬，感慨万千。

从嘉德竞拍成功到进入佳士得展览，前后不足十天，如此紧密无缺的“完美衔接”，令人不得不怀疑，这是“蓄谋已久”还是“机缘巧合”？

在展览现场，我分别就时间线问题采访了佳士得、区氏和山外，得到的答案是，展览的确是“蓄谋已久”，从一年前就开始筹备，调用的资源和展品，无论是数量还是质量，都称得上是佳士得（北京）开业以来的之精、之最。而王世襄藏书加入展览，纯属顺势而为的“意外之喜”。

买家区锦泽告诉我，这些年跟随父亲区胜春从事中式家具制作行业，已经翻烂了好几本王世襄先生的《明式家具研究》，从未想过有一天能这么近距离地接触到王老的精神殿堂。

书买到以后，他们粗粗检阅过，结果“天天有彩蛋”，“一天翻一本，天天过圣诞”，加之筹备一年的展览终于定下来于 6 月开展，王老藏书顺理成章地进入“书房”空间展览——恐怕再没有比这更合适书房的展品了。

当然，这仅仅是王老藏书的极小一部分展示。关于 4000 册图书的下一步归宿，区氏正在筹备长期的准备工作，有计划在中山打造一间中国家具主题资料馆，专门对这批藏书进行保存、展示和研究，并对公众预约开放。

三　是展览，也是美学教育

为期一周的展览期间，每天都有一场主题讲座或论坛，配合网络直播平台，线上线下同步推广美的教育。

6 月 17 日下午，佳士得在艺术中心一楼举办了一场“东西方坐具”的主题论坛，邀请了六位中式家具研究者与西方家具研究者做了精彩的分享和对话。

论坛结束后，几位嘉宾也对“退藏”聊了聊他们对展览的印象和王世襄藏书的寄望。

柯惕思 (Curtis Evarts)

美国人，著名中国古家具收藏家、学者，上海“善居”主人

我觉得这是在展示现代的生活方式。不再是整个房间都展示黄花梨家具，而是用黄花梨家具搭配古今中外各式艺术品，空间里有了木，还加入了石头、植物，还能看到金、铁、陶等元素的器物，就像中国讲究“五行”的概念一样，空间有了五行元素，就会自然变得平衡和协调。

我认识区氏有十几年了，最近还去过两三次。他们在不断研究、复制和改良明式经典家具，在行业内很受认可。这次他们和佳士得合作，以展览的方式展现传统家具的多样性，我觉得有很好的教育意义——这样的事情本来应该是政府和博物馆来做的。

我和王世襄先生是老朋友，嘉德拍卖时我有去看预展，也很关注这批书的动向。

王世襄受过很好的英文教育，他的外国朋友非常多，他的藏书里就有很多外国友人赠送给他的书，有些在图书馆里现在已经找不到了。我很高兴区氏收藏了这批书，中国需要更多他们这样的人敢做这样的事情，如果能更多地收集这样珍贵的书籍资料，对今后中国家具的研究和进步，都会是很好的事。

张金华

中国国家博物馆古家具鉴定专家，《维扬明式家具》作者，叙和堂主人

展览中西交汇，古今纵横，以功能及美学为主调，呈现当代生活场景的多元姿态，也是一种时代的趋势。

寄希区氏化私为公，并借鉴王世襄先生的治学之道，不断扩大传统木作家具的研究成果。

邓彬

江南大学设计学院教师，金缮修复研究者

我个人很喜欢这样的展览，古为今用的跨界混搭，在今天是一种潮流。毕竟我们是现代人，生活在现代空间，建筑空间也和明代很不一样了，审美习惯也是当下的，各种元素、各种风格的家具混搭融合，适应当下的潮流，也是大势所趋。

另外我觉得很有意思的一点是，明式家具能够完美融入这个环境，是因为它符合中国造物的精神：克制、谦虚、内敛，天生低调，因此能够和北欧、意大利家具和谐共处，相互拥抱。

区氏买到王世襄的藏书，我第一时间就得到了消息，当时很激动，对他们表示了祝贺。对于前辈文化学者的整体书籍收藏，之前没有人做过，区氏作为一个新家具企业能有这样的魄力，我挺意外，也不意外。因为王世襄是明式家具学科的奠基人，对这个行业的影响力太大了，从某种意义上而言，这十柜图书组成了他当年知识构成和思想智力的物质化空间。单本书籍的价值毕竟有限，像这样完整的一个知识构成物理空间太难得了，是一笔巨大的无形的财富。收藏了它们，相当于收藏了王世襄先生的书房，非常有意义。

这批图书不仅仅具有收藏价值，更有教育和研究的属性，我个人希望今后区氏能够开放给学者和有识之士，作为他们研究的一个阵地，让王世襄先生的思想能够发挥出更大的作用。

李孟苏

设计文化研究者，曾任《三联生活周刊》主任记者

著有《为生活的设计——丹麦设计的 9 堂课》一书等

我觉得这样的混搭展览挺好的，刚才看到有一个空间，左右摆着一对中式落地宫灯，中间是一张绿色沙发，前面是几何造型的亚克力茶几，这种带有冲突感的融入，感觉很棒。在今天处于极寒之下的古典家具市场，“汲古涵今——当代书房艺术”展以高规格、高质量的展品和陈设亮相佳士得，无疑给整个家居行业带来了积极正面的示范和提振作用，亦表现出了老牌家具企业区氏的底气和文化自信。

也正是处于市场寒冬时，此番展览才愈发显得抢眼和珍贵。有些机会，从来不是由市场冷或热带来的，而是人们亲手创造出来的。

“如果有新世界，一定是我们共同创造的新世界。”

An Ideal "Home" Built by Christie's (Beijing)

By Cheng Xiang

A pair of yellow *Huali (Huanghuali)* wood armchairs are placed on the left side; a soft and comfy sofa is placed in the middle for guests to sit around; a stand, a stool and a literati rock are placed in front, with two palace lanterns; a pair of shelf cabinets stand against the wall, with door panels made out of one-piece matieral, demonstrating a simple and subtle luxury.

In the centre of the tea room, there is a large tea table with flat-corner structure (corners butted with mitered corner bridle joints), accompanied by a yoke (official's hat)-back armchair with four protruding ends, with simple and clean structure; on the right corner of the tea table, there stands a round-cornered cabinet (with pedestal) and a round five-legged insence stand with cabriole legs; in front of the tea table, there is the "China Chair"produced by Danish designer Hans Wegner in 1944, which was very much inspired by Chinese Ming-styled horseshoe-back armchair, and fully took in the amenity and curvature of the armchair; on the tea table, there are dark-red enameled ceramic tea pot, tea cups, trash pot, lid stand, tea spoons, and tea tray, all kinds of tea wares.

Two black LC-2 sofas are placed next to the opposite wall, which was designed by Le Corbusier: they are made of stainless steel and leather, forming a harmonious

co-existence of the hard and the soft; beside the sofas, there hangs a photograph *Double-Mickey, No.2* executed by Chen Man during her college time.

The western and eastern styles are mixed up silently in this space where people can speak freely about the presentation of Chinese furniture in the contemporary context, the perspective on the introduction of old tradition into today's art practice, the interesting attempts on the fulfillment of the concept of living with art, the mix and clash of western and Chinese cultures… …what's more to be expected?

In this June, Christie's built a five indoor scenes in their art space in Beijing, in an attempt to construct an ideal "home".

Old is New

In 2015, Christie's restored Robert Hatfield Ellsworth's magnificent apartment in New York, directly presenting the living space of the most prestigious antique dealer to the public, and also bringing the lifestyle of this great western collector back to China.

After the featured sale of Ellsworth's collection, Christie's continued to bring the living spaces of important western collectors at previews, causing a new trend in the art community.

However, in China, with the convenience of the internet, Chinese people have obtained a higher standard and deeper yearning for better artwork, as well as better life.

From 13th to 21st of June, 2019, Christie's constructed an exquisite and ideal "home"at the art space in Beijing as the preview for their private sale.

This "home"is constructed by a living room, side room, a tea room, a prayer room and a study; by the revisit to the old tradition in the contemporary context, all the special guests had a series of meaningful discussions on lifestyle, art and design.

The exhibition started from people's daily life, bringing together classic Ming-styled furniture made by Ou's workshop, modernist furniture influenced by the philosophy of Bauhaus, Nordic design, furniture and decorations made by contemporary Italian designers, as well as artworks executed by prestigious ancient Chinese literati and famous artists, including Wu Changshuo, Wu Hufan, Wang Yuanqi, Jin Junming,etc., and contemporary works executed by Japanese artists, including Inoue Yuichi and Sugimoto Hiroshi, to create an artistic living space applied for daily life and also with superior aesthetic taste.

金高士孝章先生畫梅眞迹

Restoration of "Wang Shixiang's Study"

Most of you must still remember the recent widely spread article *An Artwork of the Most Profound Sentimental Attachment from Former Collection of Wang Shixiang (Shih-Hsing Wang)*: Wang Shixiang's private book collection (4000 volumes) had just sold at the auction orchestrated by China Guardian in this June; and the buyer—Ou's Classic Furnitre, is also invited to this exhibition. In the study of this exhibition displays dozens of books collected by Wang Shixiang (part), which has become an interesting highlight.

Some books are presented with some letters or invitation cards or newspaper clips; there leaves clear marks and comments made by Wang Shixiang; while looking at these books, one seems to be able to see his receding figure, reading and writing by the desk, which would definitely touch people's heart.

From the success bidding at the sale of China Guardian to the preview exhibition of Christie's, within no more than ten days, such tight schedule makes people wonder if this arrangement a beautiful coincident or a long-planned date?

At the exhibition, concerning the timeline, I respectively interviewed Christie's, Ou's Classic Furniture and Shan Wai (Beyond the Mountains), and got an answer that the exhibition is indeed a long planned date, which has been prepared one year ago. The exhibition calls in a large resources and top artworks, which makes it the best exhibition held in the art space in Beijing either in terms of quantity or quality. And the arrival of Wang Shixiang's private book collection (partial) is a pure surprise.

The buyer, Ou Jinze told me that he has been following his father Ou Shengchun in the classic furniture business for several years; during this time, he has read Wang Shixiang's monograph *Connoisseurship of Chinese Furniture: Ming and Early Qing Dynasties* for coutless times, yet he never imagine that one day he can own the access to Wang Shixiang's spirit hall.

After the auction, they roughly went through the collection and were surprised by various new findings every day. Since their long-planned exhibition has finally been settled to start in June, Wang Shixiang's private book collection has become an indispensable part of their art space.

Of course, it is only a tiny small part of Wang's book collection; regarding to the final destination of these books, Mr. Ou has a long-term plan in mind; they are planning to built an archive centre of classic Chinese furniture in Zhongshan City to provide better preservation, display and research on these books, and it is open to the public.

An Exhibition, Also An Aesthetic Education

During this one week exhibition, there are various theme lectures or forums held every day, along with the live streaming on the internet, to better promote the aesthetic education.

In the afternoon on 17th of June, Christie's held a theme forum "Seating Furniture in the East and West" on the ground floor; they invited six scholars, respectively specializing in traditional Chinese furniture and western furniture to share their thought and experience.

After the forum, several guests also talked about their impressions on the exhibition and good wishes for Wang Shixiang's private book collection:

Curtis Evarts

American, famous collector of classic Chinese furniture, scholar, owner of "Shan Ju"(Hillside House) in Shanghai

---

I feel that the exhibition is showcasing a lifestyle, instead of a full room of yellow *Huali (Huanghuali)* wood furniture. They applied the furniture into the scenes with various artworks from different periods and ethnics. There are wood, rock, plants, as well as gold, iron, ceramics, which perfectly illustrate the traditional concept of "balance of five elements".

I've known the family Ou for couple of years, and have recently visited them two or three times; they have been keeping to study, make the replica of the classic, and renovate the Ming-styled furniture, which makes them widely recognized in the business. This time, they worked with Christie's and showed the public the diversity of traditional furniture. I believe it is a meaningful aesthetic education which should have been done by the government or state-owned museum.

Wang Shixiang and I were old friends; during the Guardian's featured sale, I went to see the preview and surly paid high attention on the book collection. Wang received very good English education; he had many foreign friends; there are many books in his collection were given by his foreign friends as gifts; some of them already can't be found in libraries; I am very pleased that family Ou bought the whole collection. China needs more people like them who dare to do such things; if more and more books and rare manuscripts can be collected and gathered at one place, it shall be of great help for the research and development on the classic Chinese furniture.

Zhang Jinhua
Expert in the Department of Ancient Furniture of National Museum of China, writer of *Ming-style Furniture in Weiyang Area*, owner of "Xuhe Hall"

This exhibition expands the interaction between the east and the west, the past and the present, focuses on the function and aesthetics, and presents a diversified appearance of contemporary living scenes. Hope that family Ou can take the private book collection to the public and benefit more people, and to keep on the research on traditional wooden furniture in light of Wang Shixiang's guidance.

Deng Bin
Professor at the School of Design in Jiangnan University, expert in Kintsukuroi

I do like this kind of exhibition, to mix up the ancient and the present, which is a trend in the art world. After all, we are living in the 21st century; all the architectural spaces are quite different from the ones in the Ming Dynasty; we are used to the contemporary aesthetics, with the mix and blending of all kinds of elements and styles; the display and appliance of classic Chinese furniture also shall follow the current.

Another thing I found very interesting is that Ming-styled furniture can perfectly blend in the contemporary space; as they were built following the essences of Chinese philosophy: restraint, humbleness, and keeping low-profile, therefore, they can go along with Nordic and Italian furniture in harmony and peace.

I immediately got the news when Mr. Ou bought Wang Shixiang's private book collection. I was thrilled and expressed my greetings. The collection of the private books owned by a predecessor in academia was never done before Mr. Ou. I was surprised that as a new company in this business, they have such a bold courage; and I also kinda thought about it that because Wang Shixiang was the founder in the study of Ming-styled furniture, he had profound influence on this business. In a sense, this set of book collection consisted his knowledge system and is the physical reflection of his spiritual world. The single book only has limited value; yet with all these book collection, one can restore Mr. Wang's universe of knowledge, which is quite rare and unique treasure. Having them means to have bought the whole study ever owned by Wang Shixiang. It is very meaningful.

This batch of books are not only of collection value, also of educational and research values. I personally hope that Mr. Ou can open the archive to the public to benefit more people, and make it an institute of their research on classic Chinese furtniture, to pass on Wang Shixiang's spirit.

Li Mengsu

Researcher on Design Culture, ex-chief reporter of *Sanlian Life Week*, writer of the book *Design for the Life—Nine Classes of Danish Design*

I really like this mixed up style; I just saw a pair of palace lanterns placed at that corner; between them, there is a green sofa, accompanied with a geometrical tea table made of acrylic, which truly presenting a striking contrast. During the deep winter of classic Chinese furniture, this exhibition held with such high quality and broader scope at the art space of Christie's, is definitely bringing a positive energy to the business and also sets an great example to all and shows a lot cultural confidence from Ou's Classic Furniture.

It is during such a hard time, this exhibition turns out to be more meaningful. Some opportunities, are never brought by the market; they are created by people.

"If there was a new world; it must be the one we created together."

图书在版编目（CIP）数据

区氏家具 / 区胜春著. — 北京 : 中华书局, 2022.11
ISBN 978-7-101-15341-5

Ⅰ. 区… Ⅱ. 区… Ⅲ. 家具—中国—图集 Ⅳ. TS666.2-64

中国版本图书馆 CIP 数据核字（2021）第 180096 号

书　　名	区氏家具（全二册）
著　　者	区胜春
责任编辑	许旭虹　喻　济
责任印制	管　斌
装帧设计	李猛工作室
出版发行	中华书局
	（北京市丰台区太平桥西里38号 100073） http://www.zhbc.com.cn E-mail: zhbc@zhbc.com.cn
制版印刷	北京雅昌艺术印刷有限公司
版　　次	2022 年 11 月北京第 1 版 2022 年 11 月北京第 1 次印刷
规　　格	开本 787 毫米 ×1092 毫米　1 / 8 印张 59 ¾
国际书号	ISBN 978-7-101-15341-5
定　　价	1600.00 元